Handbook of Flexible Manufacturing Systems

EDITED BY

Nand K. Jha

Mechanical Engineering Department
Manhattan College
Riverdale, New York

ACADEMIC PRESS, INC.

HARCOURT BRACE JOVANOVICH, PUBLISHERS

San Diego New York Boston

London Sydney Tokyo Toronto

This book is printed on acid-free paper. ♾

Academic Press, Inc.
San Diego, California 92101

United Kingdom Edition published by
Academic Press Limited
24–28 Oval Road, London NW1 7DX

Library of Congress Cataloging-in-Publication Data

Handbook of flexible manufacturing systems / [edited by] Nand K. Jha.
p. cm.
Includes index.
ISBN 0-12-385310-9
1. Flexible manufacturing systems. I. Jha, Nand K.
TS155.6.H364 1991
670.42'7dc20 90-23926
CIP

PRINTED IN THE UNITED STATES OF AMERICA
91 92 93 94 9 8 7 6 5 4 3

Handbook of Flexible Manufacturing Systems

Contents

Contributors

Numbers in parentheses indicate the pages on which the authors' contributions begin.

Sadashiv Adiga (89), Department of Industrial Engineering and Operations Research, University of California, Berkeley, Berkeley, California 94720

Alan M. Behrens (195), Northern Natural Gas Company, Omaha, Nebraska 68102

Jim Browne (217), CIM Research Unit, University College, Galway, Republic of Ireland

F. Fred Choobineh (195), University of Nebraska at Lincoln, Lincoln, Nebraska 68588-0518

Maged Dessouky (89), Department of Industrial Engineering and Operations Research, University of California, Berkeley, Berkeley, California 94720

William D. Engelke (111), Research Institute, University of Alabama in Huntsville, Huntsville, Alabama 35899

Gary W. Fischer (61), Department of Industrial Engineering, The University of Iowa, Iowa City, Iowa 52242

Andrew Kusiak (1, 147), Department of Industrial Engineering, The University of Iowa, Iowa City, Iowa 52242

Ajit K. Patankar (275), Allied-Signal Aerospace Company, Phoenix, Arizona 85010

Vinay S. Sohoni (49), IBM Corporation, Somers, New York 10589

J. Feargal Timon (217), CIM Research Unit, University College, Galway, Republic of Ireland

H. J. Warnecke (293), University of Stuttgart and Fraunhofer Institute for Manufacturing Engineering and Automation (IPA), Stuttgart, Germany

Preface

There are a variety of terms related to factory automation, such as computer-aided manufacturing (CAM), computer-integrated manufacturing (CIM), and flexible manufacturing systems (FMSs). Still another term that is used widely in industry with regard to factory automation is CAD/CAM. In fact one can hardly find any engineering journal or conference proceedings that does not have an article on automation of factories. Even so, automated factories are not common, and it could safely be said that the necessary technology is still in its infancy. However, it must be noted that this is a rapidly developing technology and to remain competitive in the marketplace manufacturers in almost any industry must acquire it.

Factory automation is a technology concerned with the application of mechanical, electronic, and computer-based systems to operate and control production. Normally, factory automation has been grouped under three categories: fixed automation, programmable automation, and flexible automation. Fixed automation is a system in which the sequence of operation (or assembly) is fixed by equipment configuration.

In programmable automation, the production equipment is designed with the capability to change the sequence of operation to accommodate different product configurations. The operation sequence is controlled by a program, which is a set of instructions coded so that the system can read and interpret them. New programs can be prepared in order to produce new products. Examples of programmable automation include numerically controlled machine tools and industrial robots.

Computer-aided manufacturing often includes a direct numerical control (DNC) program capable of editing the manufacturing programs, downloading, and receiving feedback data from the numerical control (NC) equipment. It can prepare the schedules, batches, tools, fixtures requirement files, and production plans in a much shorter period of time than with specific packages of business systems such as the master scheduler or the manufacturing resources planner.

Flexible automation is an extension of programmable automation. Its principles are still evolving. A flexible manufacturing system is one that is capable of producing a variety of products (or parts) with virtually no time lost for changeover from one product to the next. A flexible manufacturing system typically consists of a set of cells, a material handling system that connects those cells, and service centers (e.g., material warehouse, tool room, or repair equipment). The cell is an autonomous unit that performs a specific manufacturing function (e.g., a machining center, inspection machine, or a load–unload robot).

Sometimes FMS is defined as "a set of machines in which parts are automatically transported under computer control from one machine to another for processing." In order to justify the adjective "flexible," the manufacturing system must have the ability to process a wide variety of parts or assemblies without intervention from the outside to change the system.

The computer has had a dominant impact on the development of production automation technologies. The term *computer-integrated manufacturing* is so pervasive that it is defined by CAM-I as "a series of interrelated activities and operations involving the design, material selection, planning, production, quality assurance, management, and marketing of discrete consumer and durable goods." CIM is a deliberate integration of an automated system into the process of producing a product. CIM can be considered as the logical organization of individual engineering, production, and marketing/support functions into a computer-integrated system. CIM basically deals with automating the information processing activities in manufacturing whereas automation deals with the physical activities involved in manufacturing. The growing use of computers in manufacturing systems is leading us toward the computer automated factory of the future. The term *CAD/CAM* is used interchangeably with computer-integrated manufacturing. However, CIM possesses a slightly broader meaning than CAD/CAM.

It is quite obvious that CIM is a much more advanced version of a computer automated factory and includes all FMS functions as well as the business of the company. The aim of this handbook is to present the major components of FMSs concentrating on the computer system architecture. It will also include the functions of specific cells consisting of machine tools, pallet changing and tool changing systems, part washing stations, sheet metal manufacturing cells, welding cells, assembly robots with automated hand changing and part loading facilities, and other new developments.

Flexible manufacturing systems are designed for small batch (low volume) and high variety conditions. They are designed to fill the gap between high-production transfer lines and low-production NC machines. However, it must be mentioned that only under certain circumstances are FMSs clearly justified. From literature and trade journals, it could easily be established that FMSs might be appropriate in the following situations: (1) Production of families of parts, (2) random launching of parts into the system, (3) reduced manufacturing lead times, (4) reduced in-process inventory, (5) increased machine utilization, (6) reduced direct and indirect labor, and (7) better management control.

However, designing a good FMS that will work in a particular plant with a fixed product mix is a difficult task. Most FMSs are designed with the help of simulation. Many different simulation languages are available to help in the investigation of the cells' layouts, sizes, types of equipment, and scheduling policies prior to purchase and installation. It is well known that there is a large amount of capital investment in such computer automated systems, so the justification of FMSs must also be analyzed economically.

The characteristics of U.S. FMSs have been surveyed, and it has been found that material-cutting functions are involved in about 84% of the systems and conveyer material-handling systems in about 35%. Approximate capital investment in one FMS ranges from $10 million to $25 million. FMSs with more than $10 million invested have at least 7 machining centers that are interconnected by complex material-handling systems.

This handbook covers various aspects of FMSs in different chapters. It presents in turn the components of FMSs: planning, scheduling, and control; computer control; software; simulation; databases; group technology; a practical method for FMS justification; design of flexible assembly systems; a survey of U.S. applications; and finally the unmanned factory of the future. We are grateful to the authors, who are experts in their fields, for their contributions. These contributions have made this handbook truly state of the art.

The handbook is organized by chapter as follows.

1. Andrew Kusiak details the planning, scheduling, and control involved in FMSs.
2. Vinay S. Sohoni takes an in-depth look at the computer control of FMSs, including the production control, tool control, and traffic control.
3. Gary W. Fischer describes the various software available on the market for FMSs, including those necessary for control and maintenance. This chapter looks at software from the viewpoint of market requirement.
4. Sadashiv Adiga and Maged Dessouky survey the simulation languages and real-world applications of simulation in FMS plants.
5. William D. Engelke makes a detailed analysis of databases in FMSs.
6. Andrew Kusiak writes about group technology as related to FMSs. Dr. Kusiak is an acknowledged expert in this field.
7. Alan M. Behrens and F. Fred Choobineh write about economic aspects of FMS and describe the economic conditions under which FMSs need to be installed.
8. Jim Browne and J. Feargal Timon take a hard look at flexible assembly systems (FASs) and describe various systems that they have designed personally.
9. Ajit K. Patankar writes about FMS installations in the United States. He describes some actual functioning systems of various types.
10. H. J. Warnecke describes the unmanned factory of the future. Dr. Warnecke is an internationally known and acknowledged expert in this field.

This handbook attempts to assess the current status of flexible manufacturing systems, how to effectively acquire and use this technology, and how it is going to grow in the future and eventually lead to the unmanned factory. It is comprehensive and covers all aspects of basic FMS technologies including hardware and software tools available on the market. It discusses current and projected development as well as the acquisition, evaluation, and management of systems. This handbook is intended for manufacturing managers and engineers who currently operate and/or plan to acquire an FMS.

A good deal of deliberation went into selecting the experts who have contributed to the handbook. I, as Editor, would like to acknowledge and thank all the experts who contributed to its production. It is to all these persons that I dedicate this handbook.

Nand K. Jha

Handbook of Flexible Manufacturing Systems

CHAPTER 1

Planning, Scheduling, and Control of Flexible Manufacturing Systems

Andrew Kusiak
Department of Industrial Engineering
The University of Iowa
Iowa City, Iowa

I. Introduction

One of the most difficult problems arising in flexible manufacturing systems (FMSs) is the scheduling problem. Scheduling may refer to the following subsystems of a flexible manufacturing system: fabrication, machining, and assembly.

Kusiak (1988) presented a hierarchical approach linking the machining and assembly system, where the overall FMS scheduling problem was structured as an aggregate scheduling (upper level) problem and real-time scheduling (lower level) problem. At the aggregate level the scheduling problem was modeled as

HANDBOOK OF FLEXIBLE MANUFACTURING SYSTEMS

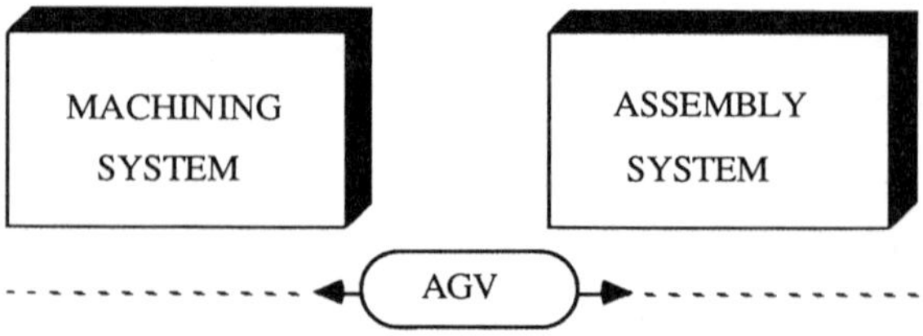

Figure 1. Structure of a manufacturing system.

the two-machine flow shop problem and solved by Johnson's algorithm (Johnson, 1954). To solve the real-time scheduling problem, a heuristic algorithm was developed.

This chapter expands the aggregate and real-time scheduling problem. Some information provided by the aggregate schedule, for example, due dates, are input to the knowledge-based scheduling system (KBSS) discussed in the last section of this chapter. To solve the real-time scheduling problem, classical and knowledge-based approaches are discussed. Readers interested in the classical scheduling theory may refer to Baker (1974) and French (1982). An extensive treatment of the knowledge-based approach to scheduling in manufacturing systems is provided in Kusiak (1990).

Consider a flexible manufacturing system that consists of a machining subsystem and an assembly subsystem. The two subsystems are linked by a material handling carrier, for example, an automated guided vehicle (AGV), as shown in Figure 1.

Consider an example product C with parts to be machined and then assembled (Figure 2). It consists of subassembly A_1, final assembly A_2, and three parts, P_1, P_2, P_3.

Parts P_1 and P_2 are to be machined before the subassembly A_1 is obtained. Assembling P_3 and A_1 results in the product C (final assembly A_2 in Figure 2).

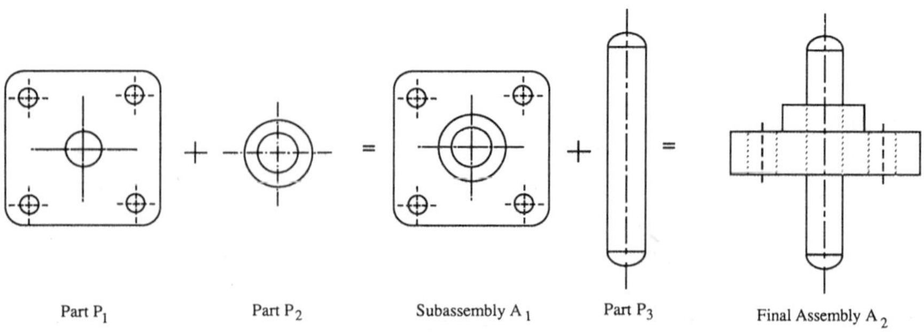

Figure 2. An example product C.

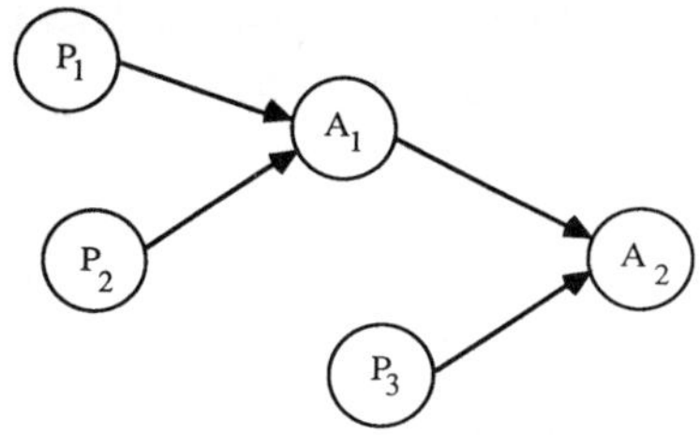

Figure 3. A digraph of the example product C in Figure 2.

The precedences among machining and assembly operations for the product can be represented by a directed graph (digraph) shown in Figure 3.

In this digraph any node of degree 1, i.e., with the number of edges incident to the node equal to 1, denotes a part; and any node of degree greater than 1 denotes a subassembly or a final product.

Another example of a digraph is shown in Figure 4(a). Without loss of generality, in this chapter, rather than representation of the digraph in Figure 4(a), the representation shown in Figure 4(b) is used. The latter representation

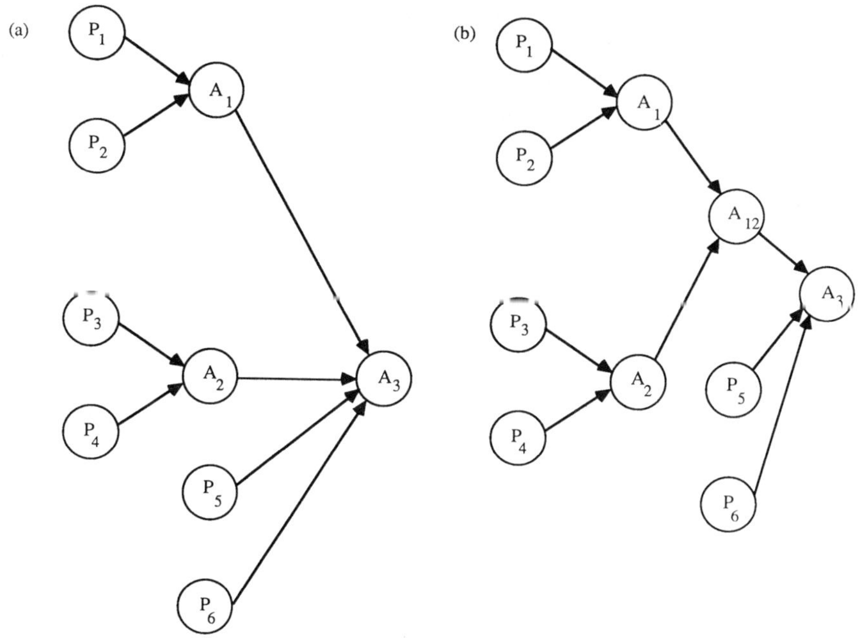

Figure 4. Two different representations of the same product.

does not allow one to assemble at a particular node more than one subassembly with any number of parts. At node A_3 in Figure 4(a), subassemblies A_1, A_2 and parts P_5, P_6 are assembled. The same subassembly A_3 has been obtained using the representation in Figure 4(b), where an additional subassembly A_{12} was inserted.

Four different aggregate scheduling problems are considered:

1. The single-product scheduling problem concerned with scheduling parts and subassemblies belonging to a single product
2. The N-product scheduling problem concerned with scheduling parts and subassemblies for N distinct products
3. The single-batch scheduling problem concerned with scheduling parts and subassemblies for a batch of n identical products.
4. The N-batch scheduling problem concerned with scheduling of parts and subassemblies for N batches of products

II. The Single-Product Scheduling Problem

Consider a digraph representation of the product which consists of a number of parts and subassemblies. In the digraph each node is labeled (a,b,c), where a is the machining time, b is the subassembly time, and c is the level of depth of the node considered. The level of depth is assigned as follows: value 0 is assigned to the root node (for example, node A_2 in Figure 3) and, working backward from the root node to the initial nodes (i.e., nodes P_1, P_2, and P_3), values of increment 1 are assigned. Digraph G from Figure 3 with labeled nodes is illustrated in Figure 5.

Before an algorithm for solving the single-product scheduling problem will be developed, a definition and two theorems are presented.

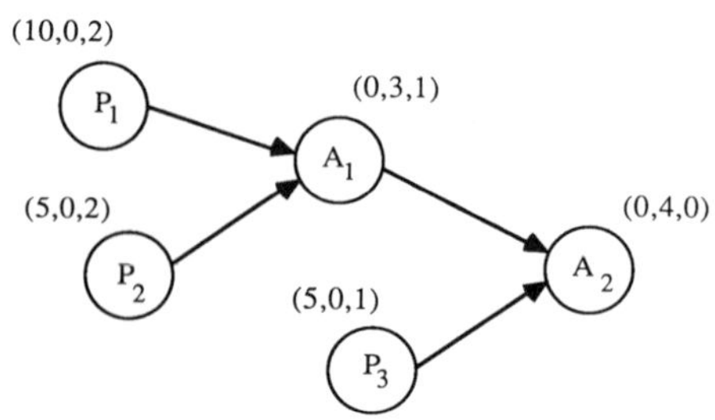

Figure 5. A digraph with labeled nodes.

Definition

A *simple digraph* G_s is a digraph in which each node of a degree greater than 1 has at most one preceding node of a degree greater than 1 [see Figure 6(a)]. Consequently, a *complex digraph* G is a digraph which is not a simple digraph [see Figure 6(b)].

Based on the preceding definition, it is obvious that any complex digraph can be decomposed into simple subdigraphs by removing a number of nodes corresponding to the final assembly or subassemblies.

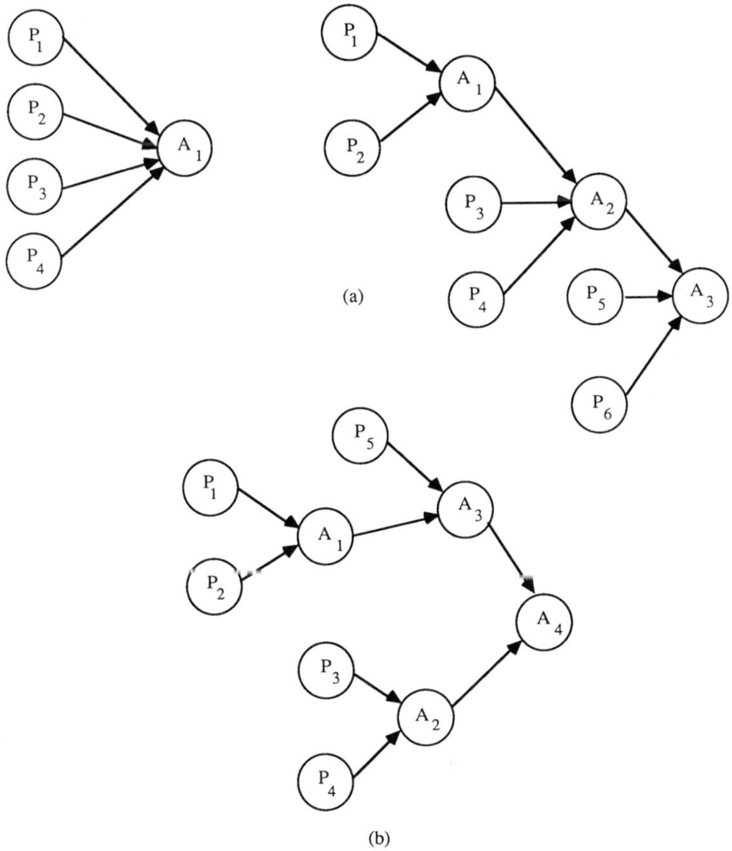

Figure 6. Examples of two types of digraphs: (a) two simple digraphs G_s and (b) complex digraph G.

Theorem 1 (see appendix for the proof)

Scheduling nodes (parts or subassemblies) of a simple digraph G_s with the maximum level of depth first (MLDF) provides the minimum makespan schedule. Theorem 1 is illustrated in Figure 7.

In Figure 7(b) the in-process idle time refers to the assembly subsystem, whereas the terminal time refers to the machining subsystem.

Theorem 2

Consider a subassembly or final product C represented by a complex digraph G and decompose it into subdigraphs $g_1, g_2, \ldots, g_t$ by removing the root node v_0 of G. Let $S(g_i)$ be the minimum makespan partial schedule associated with g_i, $i = 1, \ldots, t$. If parts and subassemblies corresponding to g_i and g_j, $i \neq j$ are preempted, then the minimum makespan schedule of product C is as follows:

$$S(C) = \{S_1(G), S_2(G), v_0\}$$

where

$$S_1(G) = [S(g_{[1]}), S(g_{[2]}, \ldots, S(g_{[k]})], \text{ for } I_{[i]} \leq T_{[i]}$$

$i = 1, \ldots, k$ is a schedule obtained using the longest in-process idle time last (LITL) rule and

$$S_2(G) = [S(g_{[k+1]}), S(g_{[k+2]}), \ldots, S(g_{[t]})], \text{ for } I_{[i]} > T_{[i]}$$

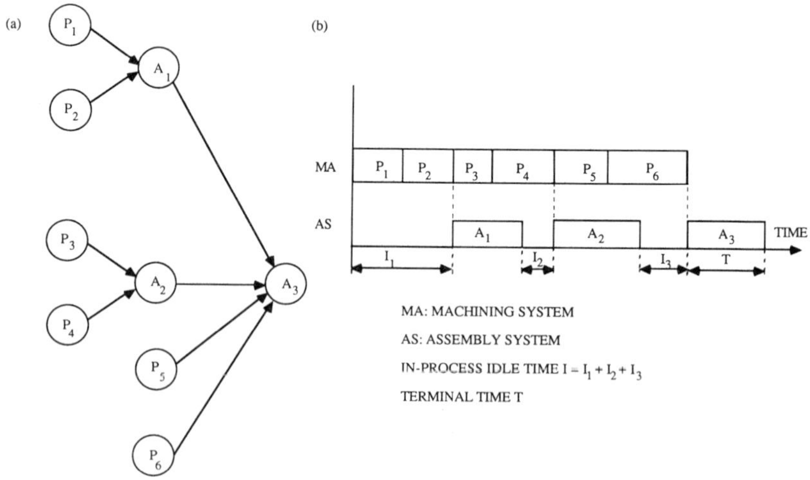

Figure 7. Application of the MLDF scheduling rule: (a) simple digraph illustrating Theorem 1 and (b) corresponding minimum makespan schedule.

$i = k + 1, k + 2, \ldots, t$ is a schedule obtained using the longest terminal time first (LTTF) rule

The proof of this theorem follows from the considerations presented in Kurisu (1976). Based on Theorems 1 and 2, Algorithm 1 is developed.

Algorithm 1 (The Single-Product Scheduling Problem)

Step 1. Label all nodes of the digraph G representing the structure of the product considered. If G is a simple digraph, use the MLDF rule to generate optimal schedule of product C, stop; otherwise, go to step 2.

Step 2. Remove root node v_0 from G and decompose it into subdigraphs g_l, $l = 1, \ldots, L$. If all g_l are simple digraphs, set $k = 0$ and go to step 3; otherwise, decompose each g_l which is not a simple digraph into a simple digraph by removing its root node. Let v_j denote a root node which has been removed, $j = 1, \ldots, J$. (Note that, for convenience, the removed nodes are numbered in the increasing order starting from the root node of G.) Set $k = J$ and go to step 3.

Step 3. Let g_{ik} denote the simple subdigraphs associated with v_k. Use the MLDF rule to generate the minimum makespan partial schedule $S(g_{ik})$ for each subdigraph g_{ik}, $i = 1, \ldots, N_k$, where N_k is the number of subdigraphs obtained after v_k has been removed.

Step 4. For each partial schedule $S(g_{ik})$ obtained in step 3, determine
(i) the in-process idle time I_{ik}
(ii) the terminal time T_{ik}, $i = 1, \ldots, N_k$

Step 5. Separate the $S(g_{ik})$ into two lists:
List 1 schedules $S(g_{ik})$ such that $I_{ik} \leqslant T_{ik}$
List 2 schedules $S(g_{ik})$ such that $I_{ik} > T_{ik}$,

$$i = 1, \ldots, N_k$$

Step 6. Use the LITL rule to generate

$S_1(g_k) = [S(g_{[1]k}), S(g_{[2]k}), \ldots, S(g_{[r]k})]$,
for $S(g_{ik})$ in list 1, $i = 1, \ldots, r$

and use the LTTF rule to generate

$S_2(g_k) = [S(g_{[r+1]k}), S(g)_{[r+2]k}), \ldots, S(g_{[t]k})]$,
for $S(g_{ik})$ in list 2, $i = r + 1, \ldots, t$; $t = N_k$

Then generate the partial schedule

$$S(g_k) = [S_1(g_k), S_2(g_k), v_k]$$

Step 7. If $v_k = v_0$, then $S(C) = S(g_0)$ is the optimal schedule, stop; otherwise, go to step 8.

Step 8. Consider $S(g_k)$ as a simple subdigraph schedule and calculate I_k and T_k. Set $k = k - 1$ and go to step 3.

Algorithm 1 is illustrated in Example 1.

Example 1

Find the minimum makespan for a product C, with the structure represented by the digraph G shown in Figure 8. The values of machining and assembly times are as follows:

Part number	1	2	3	4	5	6	7	8	9	10	11	12
Machining time	1	3	2	3	4	1	4	2	1	2	1	1
Subassembly number	1	2	3	4	5	6	7	8	9	10	—	—
Assembly time	3	1	2	3	3	4	2	2	2	3	—	—

Solution Procedure

Step 1. Since the labeled digraph G (product C) in Figure 8 is not a simple digraph, go to step 2.

Step 2. Remove the root node v_0 and nodes v_1, v_2, v_3 so that the simple subdigraphs g_{11}, g_{21}, g_{13}, g_{23}, g_{33} are obtained (Figure 9).

Step 3. Use the MLDF rule to generate the partial schedules for the subdigraphs g_{13}, g_{23}, and g_{33} associated with v_3. The Gantt chart for each subdigraph is shown in Figure 10.

Step 4. For each simple subdigraph, the in-process idle time and terminal time are

$$I_{13} = 5,\ T_{13} = 3;\ I_{23} = 3,\ T_{23} = 4;\ I_{33} = 3,\ T_{33} = 2$$

Step 5. Since

$$I_{13} > T_{13},\ I_{23} < T_{23},\ I_{33} > T_{33}$$

subdigraph g_{23} is placed in list 1; g_{13} and g_{33} are placed in list 2.

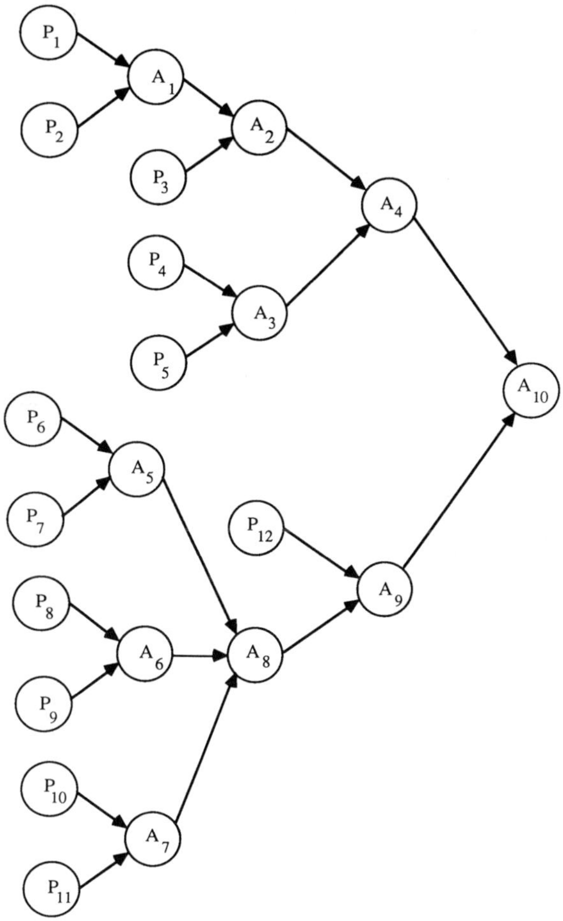

Figure 8. Product C structure.

Step 6. Use the LITL rule to generate

$$S_1(g_3) = [(P_8, P_9, P_6)]$$

and use the LTTF rule to generate

$$S_2(g_3) = [(P_6,\ P_7,\ A_5),\ (P_{10},\ P_{11},\ A_7)]$$

Then the partial schedule associated with v_3 is

$$\begin{aligned} S(g_3) &= [S_1(g_3),\ S_2(g_3),\ v_3] \\ &= [(P_8,\ P_9,\ A_6),\ (P_6,\ P_7,\ A_5),\ (P_{10},\ P_{11},\ A_7),\ P_8] \end{aligned}$$

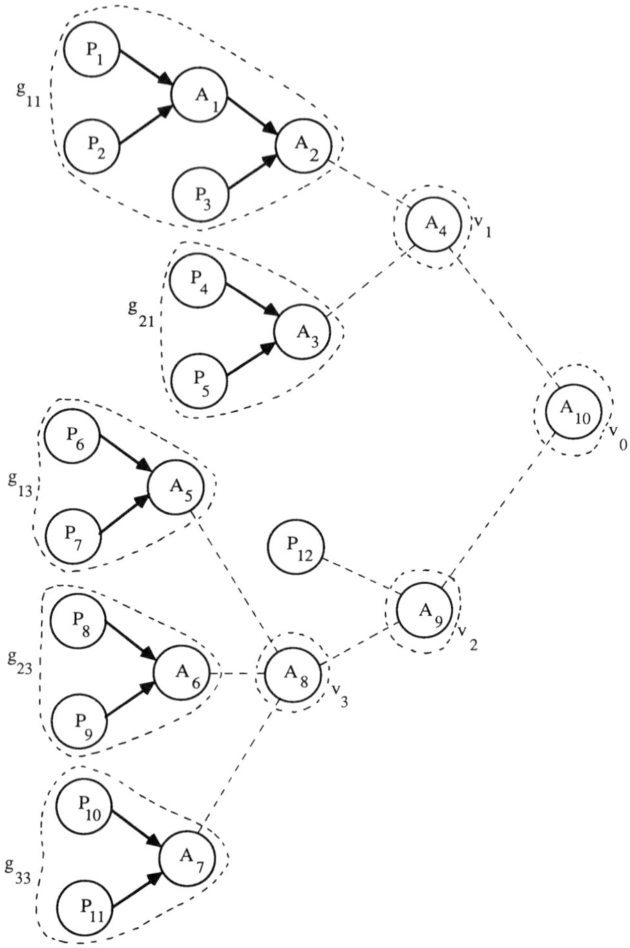

Figure 9. The decomposed digraph G of Figure 8.

The Gantt chart of the preceding partial schedule is shown in Figure 11.

Step 7. Since $v_3 \neq v_0$, go to step 8.

Step 8. For partial schedule $S(g_3)$ calculate

$$I_3 = 4, T_3 = 4$$

Set $k = 3 - 1 = 2$, go to step 3.

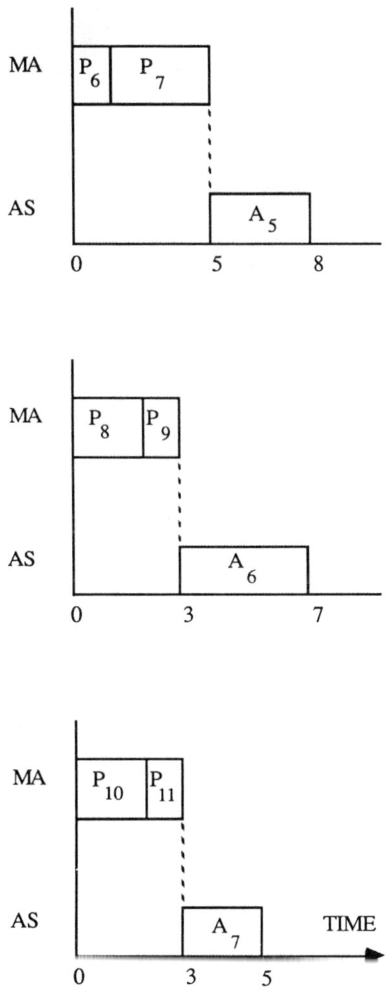

Figure 10. Gantt charts for the partial schedules of g_{13}, g_{23}, and g_{33} in Figure 9.

In step 3 (the second iteration), a simple subdigraph associated with v_2 is obtained. Using the MLDF rule, a partial schedule illustrated with the Gantt chart is shown in Figure 12.

The following values for in-process idle time and terminal time are calculated

$$I_2 = 4, T_2 = 5$$

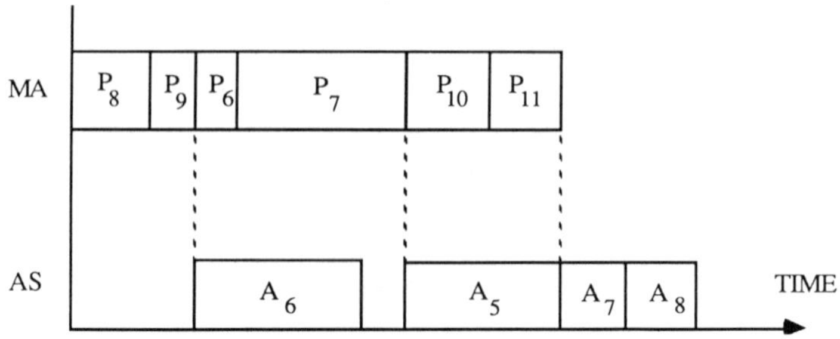

Figure 11. Partial schedule corresponding to the subdigraph with the node v_3 in Figure 9.

The same process is applied to v_1, and the Gantt chart of each partial schedule is shown in Figure 13. The corresponding values of in-process idle time and terminal time are calculated

$$I_1 = 9, T_1 = 5$$

In the last iteration, the above sequencing process is applied to v_0; and since $(I_2 < T_2)$, g_2 is placed in list 1, $(I_1 > T_1)$, g_1 is placed in list 2.

The minimum makespan schedule for product C (Figure 8) is

$$\begin{aligned} S(C) &= (S_1(g_0), S_2(g_0), v_0) \\ &= \{[(P_8, P_9, A_6), (P_6, P_7, A_5), (P_{10}, P_{11}, A_7), A_8, P_{12}, A_9] \\ &\quad [(P_1, P_2, A_1), P_3, A_2, (P_4, P_5, A_3), A_4], A_{10}\} \end{aligned}$$

The Gantt chart of the minimum makespan schedule for product C is shown in Figure 14.

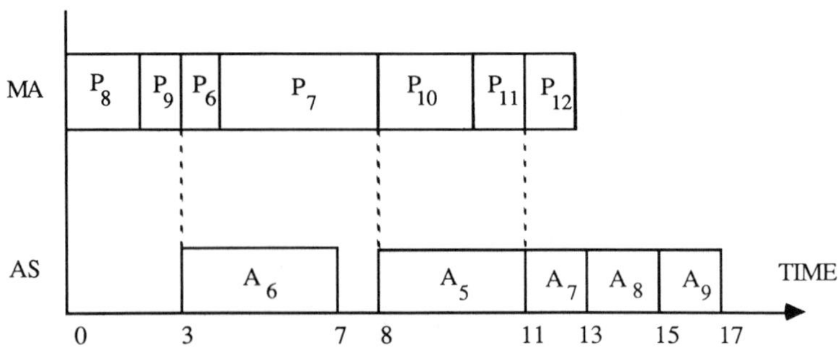

Figure 12. Partial schedule corresponding to the subdigraph with the node v_2 in Figure 9.

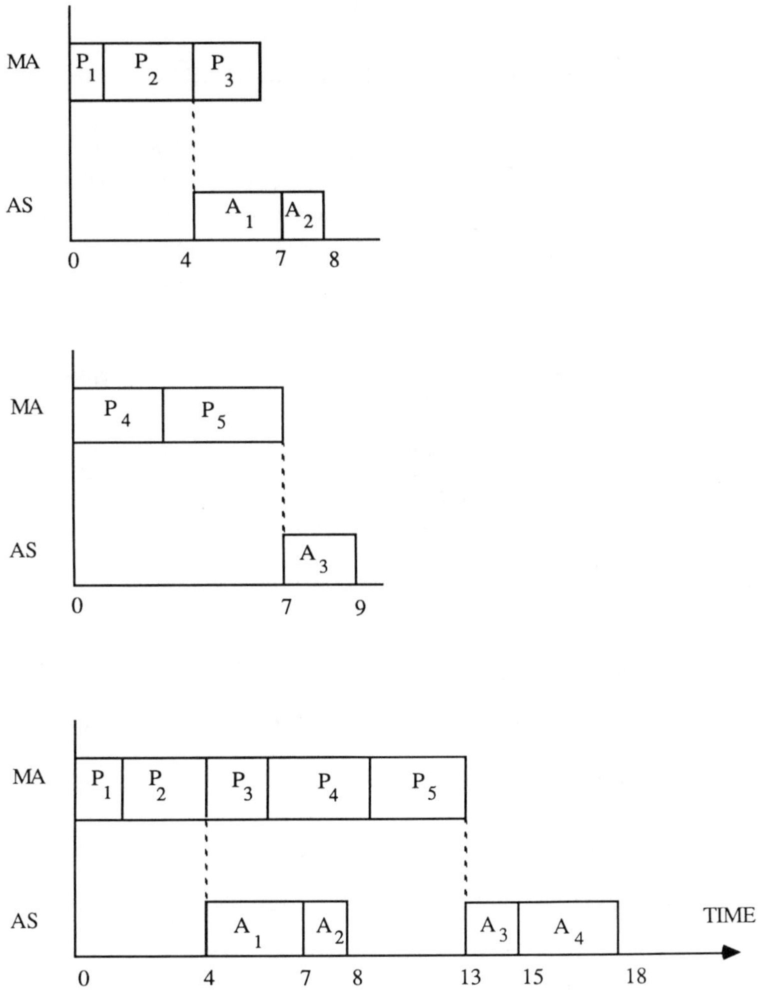

Figure 13. Partial schedules corresponding to the subdigraph with the node v_1 in Figure 9.

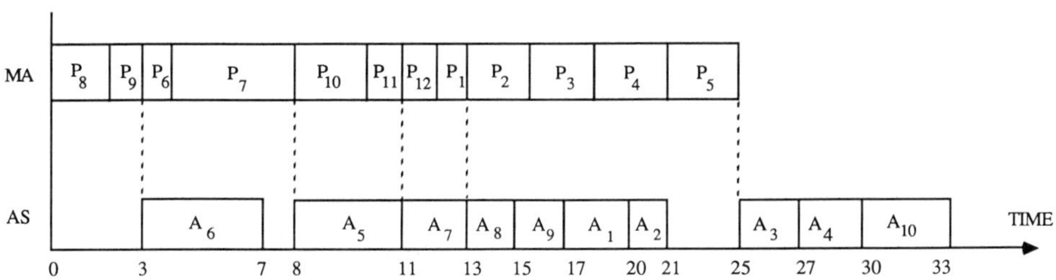

Figure 14. Gantt chart of the minimum makespan schedule.

III. The N-Product Scheduling Problem

In this section, the scheduling problem for N distinct products, each in quantity of one, is considered. To solve this problem, a "product-by-product" policy is used. The product-by-product policy assumes that the N product scheduling problem can be decomposed into N single-product scheduling problems. The algorithm for scheduling of N products $C_1, C_2, \ldots, C_N$ is presented below.

Algorithm 2 (The N-Product Scheduling Problem)

Step 1. Using Algorithm 1, determine the optimal schedule $S(C_i)$ for each product C_i, $i = 1, \ldots, N$.

Step 2. Separate all $S(C_i)$ into the following two lists:

List 1: including $S(C_i)$ such that $I_i \leqslant T_i$, $i = 1, \ldots, k$
List 2: including $S(C_i)$ such that $I_i > T_i$, $i = k + 1, \ldots, N$

Step 3. For the schedules in list 1, develop the LITL schedule:

$$S_1(NC) = \{S(C_{[1]}), S(C_{[2]}), \ldots, S(C_{[k]})\}$$

For the schedules in list 2, develop the LTTF schedule:

$$S_2(NC) = \{S(C_{[k+1]}), S(C_{[k+2]}), \ldots, S(C_{[N]})\}$$

Step 4. Generate final schedule $S(NC) = \{S_1(NC), S_2(NC)\}$

Theorem 3

The schedule $S(NC) = \{S_1(NC), S_2(NC)\}$ generated by Algorithm 2 is the minimum makespan schedule of the N-product scheduling problem.

The proof of Theorem 3 follows the results presented in Kurisu (1976), except that the subdigraph g_i is replaced with the digraph representing product C_i.

Algorithm 2 is illustrated in Example 2.

Example 2

Consider an $N = 4$ product-scheduling problem with each product structure shown in Figure 15. For simplicity, assume that the structure of each product is represented by a simple digraph.

Step 1. Using Algorithm 1, the following schedules (illustrated in Figure 16) are obtained:

Product C_1: $S(C_1) = \{P_1, P_2, A_1\}$

Product C_2: $S(C_2) = \{(P_3, P_4, A_2), P_5, A_3\}$

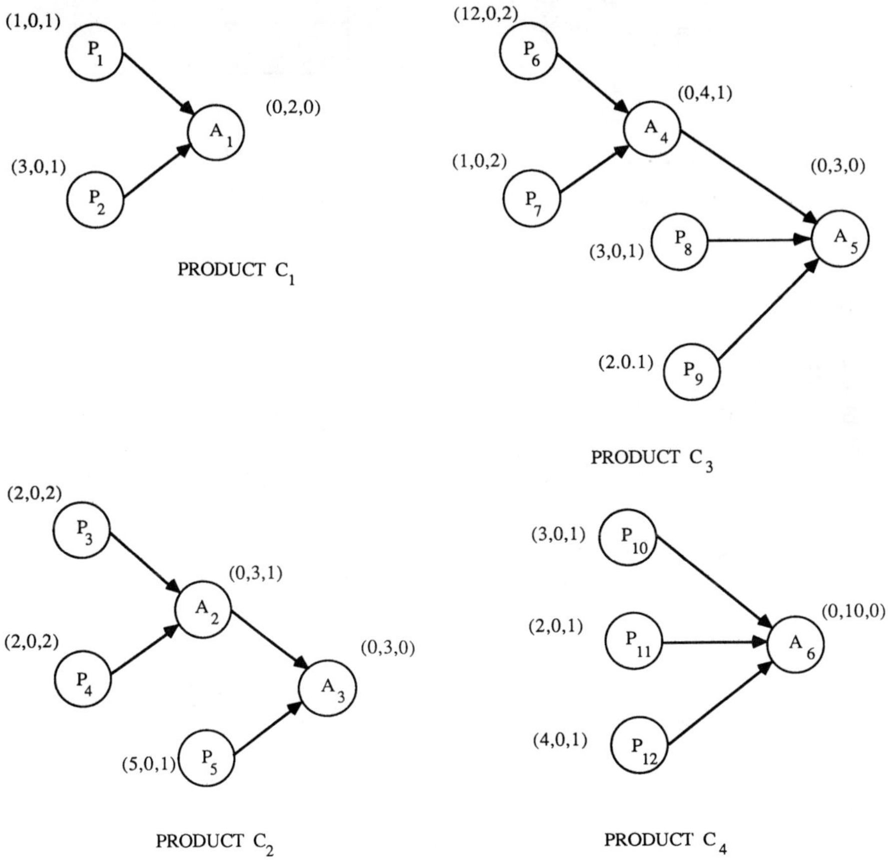

Figure 15. Structure of products C_1, C_2, C_3, and C_4.

Product C_3: $S(C_3) = \{(P_6, P_7, A_4), P_8, P_9, A_5\}$

Product C_4: $S(C_4) = \{P_{10}, P_{11}, P_{12}, A_6\}$

Step 2. From the Gantt charts in Figure 16, the following data are obtained:

Product	I_i	T_i	List number
C_1	4	2	2
C_2	6	3	2
C_3	18	3	2
C_4	9	10	1

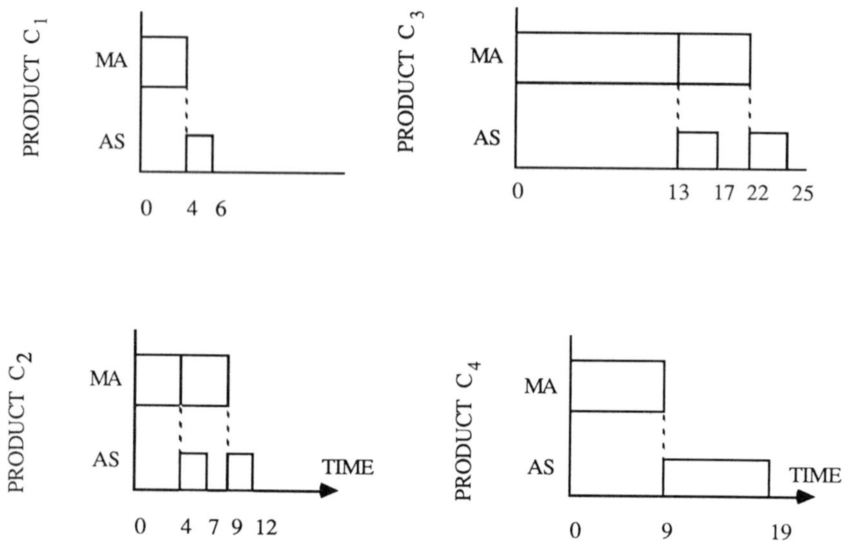

Figure 16. Schedules for the products in Figure 15.

Step 3. Using the LITL and LTTF rules results in the following schedules:

$$S_1(NC) = \{C_4\}$$

$$S_2(NC) = \{C_3, C_2, C_4\}$$

Step 4. The optimal schedule is as follows:

$$S(NC) = \{[(P_{10}, P_{11}, P_{12}, A_6)], [(P_6, P_7, A_4), P_8, P_9, A_5], [(P_3, P_4, A_2), P_5, A_3], [P_1, P_2, A_1]\}$$

IV. The Single-Batch Scheduling Problem

The algorithm for solving the single-batch scheduling problem is based on Algorithm 1. The single-product schedule is repeated *n* times, where *n* is the number of products in a batch.

Algorithm 3 (The Single-Batch Scheduling Problem)

Step 1. Using Algorithm 1, determine the schedule $S(C)$ for the single-product scheduling problem.

Step 2. Determine the schedule $S(B)$ for the single-batch scheduling problem by repeating n times the schedule $S(C)$.

The above algorithm is illustrated in Example 3.

Example 3

Develop the optimal schedule $S(B)$ for $n = 5$ identical products based on the following data:

Part or subassembly	P_1	P_2	P_3	A_1	A_2
Machining time	10	5	6	—	—
Assembly time	—	—	—	4	3

and precedence constraints:

$$P_1, P_2 \rightarrow A_1$$

$$P_3, A_1 \rightarrow A_2$$

Step 1. Solving the single-product scheduling problem, the following schedule is obtained:

$$S(C) = \{(P_1, P_2, A_1), P_3, A_2\}$$

Step 2. The optimal schedule $S(B)$ is generated as follows:

$$\begin{aligned} S(B) = \{&[(P_1, P_2, A_1), P_3, A_2], [(P_1, P_2, A_1), P_3, A_2], \\ &[(P_1, P_2, A_1), P_3, A_2], [(P_1, P_2, A_1), P_3, A_2], \\ &[(P_1, P_2, A_1), P_3, A_2]\} \end{aligned}$$

V. The N-Batch Scheduling Problem

In this section, an algorithm for solving the N-batch scheduling problem is developed. The N-batch scheduling problem can be decomposed into N single-batch scheduling problems. To solve this problem, the following algorithm is used.

Algorithm 4 (The N-Batch Scheduling Problem)

Step 1. Using Algorithm 2, determine the schedule $S(NC)$ for the N-product scheduling problem.

Step 2. Determine the schedule $S(NB)$ for the N-batch scheduling problem by repeating n_i times each schedule $S(C_i)$ in $S(NC)$, where n_i is a batch size of product C_i.

The above algorithm is illustrated in Example 4.

Example 4

Develop the optimal schedule for the $N = 4$ batch scheduling problem. The batch sizes of each product are as follows:

Product	Batch size
C_1	2
C_2	2
C_3	3
C_4	2

The structure of each product is shown in Figure 15.

Step 1. Using Algorithm 2 results in the following schedule (see Example 2):

$$S(NC) = \{C_4, C_3, C_2, C_1\}$$

Step 2. The schedule $S(NB)$ is obtained as follows:

$$S(NB) = \{[C_4, C_4], [C_3, C_3], [C_2, C_2, C_2], [C_1, C_1]\}$$

The aggregate schedules generated by the algorithms presented can be used by the knowledge-based scheduling system (KBSS) discussed later in the chapter. The aggregate schedules impose due dates that may be input to the scheduling algorithm embedded into the KBSS. Another possibility is to incorporate the data provided by the aggregate schedules into the priority rules of the KBSS.

VI. Scheduling Flexible Forging Machine

In this section, a single-machine scheduling problem with changeover costs and precedence constraints is considered. The problem is discussed using an example of a flexible forging machine.

One of the most frequently emphasized aspects of flexible manufacturing systems (FMSs) deals with changeover costs. Very often FMSs are modeled as systems with virtually no changeover costs. An attempt will be made to show that changeover costs exist in FMSs, and their reduction is a matter of a planning methodology. An FMS, if not properly planned, may in fact result in unnecessarily large changeover costs. Rather than a large-scale flexible manufacturing system, a flexible forging machine (FFM) is considered.

The flexible forging machine in Figure 17 is computer controlled and has the following four automatic options:

1. Part loading
2. Chucks rotation and horizontal movement
3. Tool magazine rotation and oscillation
4. Part loading/unloading

The flexible forging machine considered was developed by Ishikawajima-Harima Heavy Industries Co. in Japan. Each of the two identical tool magazines visible in Figure 17 can handle eight tools. Tools are heavy and it takes a considerable amount of time to change them. This will have some implications on the modeling and solution approach discussed in this section.

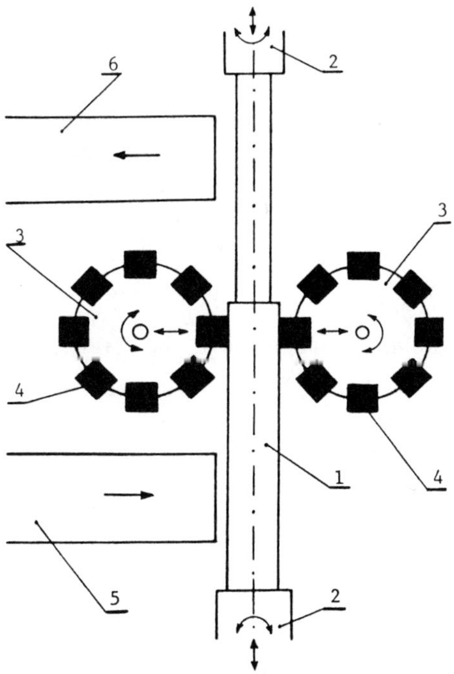

Figure 17. A schematic top-down view of a flexible forging machine (FFM).

A. FEATURES OF THE FLEXIBLE FORGING MACHINE SCHEDULING PROBLEM

1. Changeover Costs

The idea behind the FMS concept is to reduce or preferably eliminate changeover costs. Due to the progress in FMS hardware and advancement in the planning methodology, it has been possible to reduce these costs, but not to eliminate them entirely. It will be shown that, in the flexible forging machine discussed, changeover costs occur. Reduction of these costs is a matter of the flexible forging machine scheduling. There are the following four sources of changeover costs in the flexible forging machine shown in Figure 17:

1. Parts
2. NC (numerical control) programs
3. Chucks holding parts
4. Tools

The changeover cost imposed by loading and unloading of parts is dictated by the hardware design and its reduction is beyond operational control. The NC program is being changed during the loading of a new part and the unloading of the already forged part. The NC program changeover time is only a small fraction of the part loading and unloading time. The chucks are capable of holding many different parts and they are seldom changed. Tool changeover imposes significant costs, and this issue is of main concern for the optimization model presented later in this section. Moreover, these costs are sequence-dependent. Before a new part (or a batch of parts) is forged, in general, it is necessary to change some tools stored in the tool magazines.

2. Precedence Constraints

Classical scheduling problems might be formulated with two types of constraints:

1. Due dates
2. Precedences

The precedence constraints are more appropriate than the due date constraints in modeling of the flexible forging machine scheduling problem. The flexible forging machine is to be integrated with the machining and assembly systems. Often, the final products are assembled in a sequence imposed by the market demand. In this case, in order to reduce a volume of in-process inventory, it is desirable to perform the machining and forging process in a sequence imposed by the assembly process. If this sequence is not preserved, a delay in product delivery may occur.

B. MODELING THE FLEXIBLE FORGING MACHINE SCHEDULING PROBLEM

Based on the previous consideration, one can identify the underlying flexible forging machine scheduling problem as the single-machine scheduling problem with sequence-dependent changeover costs and precedence constraints. It is known from the scheduling theory (see, e.g., Baker, 1974, p. 94) that the single-machine scheduling problem with sequence-dependent changeover costs is equivalent to the traveling salesman problem (TSP).

One of the most frequently encountered formulations of the traveling salesman problem is a mixed-integer programming formulation presented below.

Denote

n = number of cities (operations in the FFM scheduling problem)

c_{ij} = traveling (setup) cost from city (operation) i to city (operation) j

u_i = nonnegative variable

$$x_{ij} = \begin{cases} 1 \text{ if city (operation) } i \text{ immediately precedes city (operation) } j \\ 0 \text{ otherwise} \end{cases}$$

The objective function of the traveling salesman problem minimizes the total travel (setup) cost.

$$\min \sum_{i=1}^{n} \sum_{j=1}^{n} c_{ij}\, x_{ij} \tag{1}$$

$$\text{s.t.} \sum_{i=1}^{n} x_{ij} = 1 \qquad j = 1, \ldots, n \tag{2}$$

$$\sum_{i=1}^{n} x_{ij} = 1 \qquad i = 1, \ldots, n \tag{3}$$

$$u_i = u_j + n x_{ij} \leqslant n - 1 \qquad i = 2, \ldots, n;\ j = 2, \ldots, n;\ i \neq j \tag{4}$$

$$x_{ij} = 0, 1 \qquad i, j = 1, \ldots, n \tag{5}$$

$$u_i \geqslant 0 \qquad i = 1, \ldots, n \tag{6}$$

Constraint (2) ensures that in a given tour only one city immediately precedes city j. Constraint (3) imposes that city i is followed by exactly one city. The formulation (1), (2), (3), and (5) is known in the operations research literature as the assignment problem. Constraint (4) eliminates subtours generated by solving the assignment problem (1)–(3). Integrality is imposed by constraint (5). Constraint (6) ensures nonnegativity.

To illustrate application of the formulation (1)–(6) for solving the forging machine scheduling problem, consider the following example.

Example 5

Find the optimal schedule for six parts. Each part is to visit a flexible forging machine only once. The matrix of sequence dependent changeover costs is as follows:

		Part number					
		1	2	3	4	5	6
Part number	1	∞	7	3	12	5	8
	2	4	∞	2	10	9	3
	3	6	7	∞	11	1	7
	4	7	3	1	∞	8	8
	5	2	10	2	7	∞	3
	6	4	11	7	6	3	∞

Solving the formulation (1)–(6) by a standard computer code suitable for solving linear 0–1 programming problems (e.g., LINDO, Schrage, 1984) or a specialized algorithm (see, e.g., Lawler *et al.*, 1985) the following optimal solution is obtained:

$$x_{12} = x_{26} = x_{64} = x_{43} = x_{35} = x_{51} = 1$$

which can be also expressed as (1, 2, 6, 4, 3, 5, 1). The cost of the optimal solution is 20.

The flexible forging machine scheduling problem can be described as the TSP problem with precedence constraints.

So far, not too many operations research models have been applied to manufacturing systems. In our view there are two basic obstacles:

1. Models formulated have been too complex
2. Algorithms for solving these models have been difficult to implement

The approach presented in this section avoids these obstacles. First, we do not consider the standard formulation of the TSP [e.g., (1)–(6)], which become cumbersome if the precedence constraints are added. Instead, a network flow formulation is presented. In this formulation the precedence constraints are expressed very easily. In addition, a good feasible solution to the network formulation can be generated by a standard linear programming (LP) code. To obtain the optimal solution, a branch and bound search will be applied.

C. TWO-COMMODITY NETWORK FLOW FORMULATION

As mentioned earlier, the sequence-dependent changeover costs lead to a traveling salesman problem with n cities. In the model discussed, the cities 1, 2, . . . , n correspond to parts (or batches of parts) $P_1, P_2, \ldots, P_n$ to be scheduled. The distance between the cities i and j are to be replaced by the changeover costs $c_{ij} = c(P_i, P_j)$ from part P_i to P_j. The traveling salesman tour, which is simply a permutation of 1, 2, . . . , n indicates directly the order of processing the parts. The precedence constraints are of the form "part P_u before part P_v" or, symbolically $P_u \rightarrow P_v$ for certain combinations of parts P_u and P_v. To model the scheduling problem, the approach given in Finke *et al.* (1984) is followed. The network flow of two distinct commodities is used to characterize the traveling salesman tours. Consider the TSP with $n = 6$ cities and the tour (1, 4, 2, 3, 6, 5). There are two commodities, p and q, given in the network. At a selected starting point s (here $s = 1$), $(n - 1)$ units of the commodity p are available, one unit for each node i, $i \neq s$. Complementary to p, commodity q has one unit at each i, $i \neq s$, and $(n - 1)$ units are required at s. Figure 18 displays the feasible flow for the tour in our example (positive numbers are supplies and negative numbers are demands).

In general, we require a feasible flow x^p_{ij} with respect to commodity p:

$$\sum_j x^p_{ij} - \sum_j x^p_{ji} = \begin{cases} n - 1 & \text{for } i = s \\ -1 & \text{elsewhere} \end{cases} \tag{7}$$

$$x^p_{ij} \geq 0 \qquad \text{for all } (i,j)$$

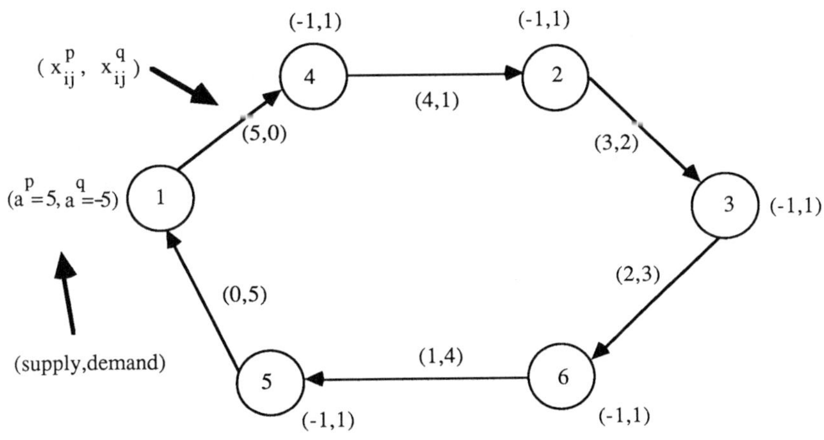

Figure 18. Feasible two-commodity flow (x^p_{ij}, x^q_{ij}) in a traveling salesman tour.

Similarly, we have for commodity q

$$\sum_j x^q_{ij} - \sum_j x^q_{ji} = \begin{cases} -(n-1) & \text{for } i = s \\ +1 & \text{elsewhere} \end{cases} \tag{8}$$

$$x^q_{ij} \geqslant 0 \qquad \text{for all } (i,j)$$

Looking at Figure 18, one can notice that the combined total $x^p_{ij} + x^q_{ij}$ is equal to $(n - 1)$ for each arc of the tour. This may be expressed as follows:

$$\sum_j (x^p_{ij} + x^q_{ij}) = n - 1 \qquad \text{for all } i \tag{9}$$

$$x^p_{ij} + x^q_{ij} = 0 \text{ or } (n-1) \qquad \text{for all } (i, j) \tag{10}$$

Constraints (7) to (10) characterize precisely all Hamiltonian tours. (A Hamiltonian tour is a loop including all the vertices of a network.) In fact, constraints (9) and (10) imply that there is exactly one arc leaving each node i and carrying a combined total flow of $(n - 1)$ units. On the other hand, there must be a flow-carrying path from s to every other node (following the commodity p) and a path from an arbitrary node i to s (following q). It is easily seen that the only possible configuration is a tour.

Since $(x^p_j + x^q_{ij})/n - 1 = 0$ or 1, we obtain a linear expression for the total cost which has to be minimized:

$$\min 1/(n-1) \sum_i \sum_j c_{ij} (x^p_{ij} + x^q_{ij}) \tag{11}$$

Let us now discuss the precedence constraints. Choose the part P_s so that there are no preceding parts and consider only the commodity p. The total amount of commodity p leaving a point is decreasing along the tour (see Figure 18). Therefore, it is required that

$$\sum_j x^p_{uj} - \sum_j x^q_{vj} \geqslant 1 \qquad \text{for } P_u \rightarrow P_v \ (v \neq s) \tag{12}$$

Replacing constraint (10) by

$$x^p_{ij} + x^q_{ij} = (n-1)\, x_{ij} \text{ with } x_{ij} = 0 \text{ or } 1 \qquad \text{for all } (i, j) \tag{13}$$

yields a mixed-integer programming formulation.

Example 6

For a given time horizon, assume that five parts are to be forged on the flexible forging machine in Figure 17. Each part requires a set of eight tools as indicated in Table 1. For $i = 1, 2, \ldots, 5$ and $k = 1, 2, \ldots, 22$, set

$$x_{ij} = \begin{cases} 1 \text{ if tool } t_k \text{ is used for part } P_i \\ 0 \text{ otherwise} \end{cases}$$

TABLE 1
Tools Required for Forging Five Parts

Tool number	Part number 1	2	3	4	5
1				1	
2	1		1	1	1
3		1			
4	1		1	1	1
5		1			
6	1		1		1
7				1	
8	1		1		1
9		1			1
10				1	
11	1		1		
12				1	
13	1				
14	1			1	
15		1			1
16			1	1	
17		1			
18			1		1
19	1	1			
20			1		
21		1			
22		1			1

For any two parts P_i and P_j, define the Hamming distance

$$d_{ij} = \sum_{k=1}^{22} \delta\,(x_{ij},\, x_{jk}) \tag{14}$$

where

$$\delta\,(x_{ik},\, x_{jk}) = \begin{cases} 1 \text{ if } x_{ij} \neq x_{jk} \\ 0 \text{ otherwise.} \end{cases}$$

The following matrix D of Hamming distances is obtained for the data in Table 1.

$$D = [d_{ij}] = \begin{array}{c} \\ 1 \\ 2 \\ 3 \\ 4 \\ 5 \end{array} \begin{array}{c} \begin{array}{ccccc} 1 & 2 & 3 & 4 & 5 \end{array} \\ \begin{bmatrix} \infty & 14 & 6 & 10 & 8 \\ 14 & \infty & 16 & 16 & 10 \\ 6 & 16 & \infty & 10 & 6 \\ 10 & 16 & 10 & \infty & 12 \\ 8 & 10 & 6 & 12 & \infty \end{bmatrix} \end{array} \tag{15}$$

The costs c_{ij} in the objective function (11) are set equal to the distances d_{ij}.

For convenience in modeling, we introduce a dummy batch (number $s = 0$) and a row and column vector of zero changeover costs. The dummy batch indicates the initial and final state of the tool magazine.

$$D = [d_{ij}] = \begin{array}{c} \\ 0 \\ 1 \\ 2 \\ 3 \\ 4 \\ 5 \end{array} \begin{array}{c} \begin{array}{cccccc} 0 & 1 & 2 & 3 & 4 & 5 \end{array} \\ \begin{bmatrix} \infty & 0 & 0 & 0 & 0 & 0 \\ 0 & \infty & 14 & 6 & 10 & 8 \\ 0 & 14 & \infty & 16 & 6 & 10 \\ 0 & 6 & 16 & \infty & 10 & 6 \\ 0 & 10 & 16 & 10 & \infty & 12 \\ 0 & 8 & 10 & 6 & 12 & \infty \end{bmatrix} \end{array} \qquad (16)$$

Let us impose the precedence constraints illustrated in Figure 19.

The mixed-integer programming problem with objective function (11) and constraints (7), (8), (9), (12), and (13) can be solved by any commercial computer code. Solving the problem with LINDO (Schrage, 1984) for the data in matrix (16) and the precedence constraints in Figure 19 produces the following solution:

$$x_{02} = 1,\ x_{25} = 1,\ x_{51} = 1,\ x_{13} = 1,\ x_{34} = 1,\ x_{40} = 1 \qquad (17)$$

which can be expressed as the schedule (0, 2, 5, 1, 3, 4, 0).

Alternatively, the problem (7)–(12) can be solved by the following branch and bound algorithm.

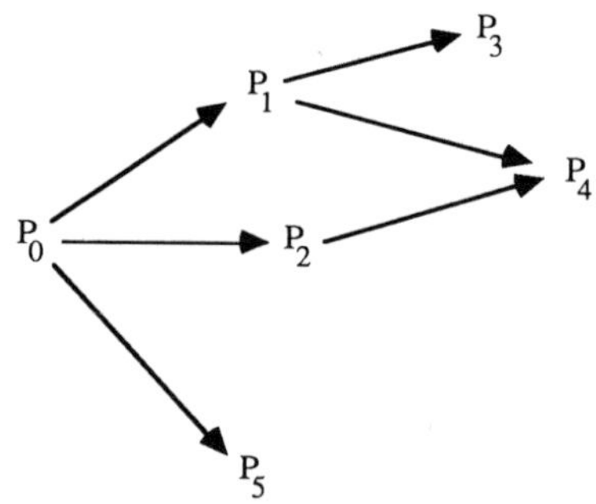

Figure 19. Precedence constraints.

Algorithm 5 (Kusiak and Finke, 1987)

Step 1. Relax constraint (10) and consider the relaxed problem P [expressions (7), (8), (9), (11), and (12)].

Step 2. Solve the linear program P. The optimal value L is a lower bound for the optimal feasible tour (feasible with respect to the precedence constraints). Concentrate on the flow graph F, characterized by the pairs of parts (or batches) (P_i, P_j) for which the optimal values x_{ij}^p and x_{ij}^q are strictly positive. If the size $|F|$ is n, then a tour and hence the solution is found. If $|F| > n$, try to detect a feasible tour in F by means of the heuristic proposed by Martello (1983). This yields an upper bound U ($U = \infty$) whenever no feasible tour exists).

Step 3. Establish a branch and bound tree and perform steps 1 and 2 at every node. The branching strategy is as follows: Select a part P_i with the maximum number of leaving arcs in the corresponding flow graph F. Branch on the arc (i, j) with the smallest cost, that is, the "left" branch uses (i, j) as the only arc leaving part P_i, and the "right" branch disallows this arc, denoted by $(\overline{i, j})$. These conditions are easily implemented using appropriate penalty costs. The

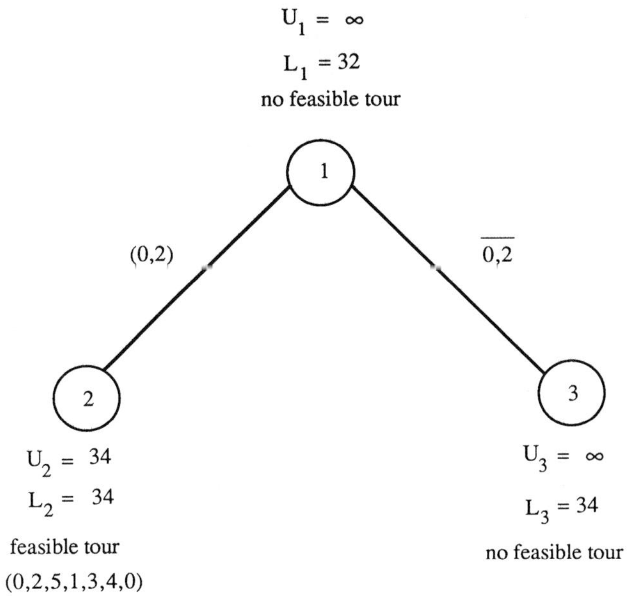

Figure 20. Branch and bound tree of Example 6.

complete branch and bound tree for the data in matrix (16) and precedence constraints in Figure 19 consists of only three nodes and is shown in Figure 20. The optimal sequence is (0, 2, 5, 1, 3, 4, 0) with the corresponding cost 34.

VII. Modeling the Problem of Scheduling n Operations on m Machines

One of the approaches to scheduling is to model the scheduling problem and then solve it using a commercial computer code or specialized algorithm. In this section a new mixed-integer programming model for the problem of scheduling n operations with precedence constraints on m machines is presented. The scheduling problem considered is a generalization of the scheduling problem presented in Baker (1974).

To present the scheduling model the following notation is introduced:

n = number of parts

m = number of machines

R_i = set of pairs of operations $[k, l]$ for part P_i, where k precedes l, $i = l, \ldots, n$

Q_i = set of pairs of operations $[k, l]$ for part P_i, where k and l can be performed in any order, $i = l, \ldots, n$

I_i = set of operations without precedence constraints, $i = l, \ldots, n$

N_p = set of operations to be performed on machine p, $p = l, \ldots, m$

n_i = number of operations in part P_i, $i = l, \ldots, n$

t_{ik} = processing time of operation k of part P_i, $k = l, \ldots, n$, $i = l, \ldots, n$

M = an arbitrary large positive number

x_{ik} = completion time of operation k on part P_i, $k = l, \ldots, n$, $i = l, \ldots, n$

$x_{(i)}$ = completion time of the root operation of part P_i, $i = l, \ldots, n$

$$y_{kl} = \begin{cases} 1, & \text{if operation } k \text{ precedes operation } l, \\ 0, & \text{otherwise,} \end{cases}$$

$k = l, \ldots, n_i$, $l = 1, \ldots, n_i$, $i = 1, \ldots, n$

The objective of the scheduling model is to minimize the total completion time of all parts.

$$\min \sum_{i=1}^{n} x_{(i)} \qquad (18)$$

s. t. $x_{il} - x_{ik} \geqslant t_{il}$ for all $[k, l] \in R_i$, for all i (19)

$x_{il} - x_{ik} + M(1 - y_{kl}) \geqslant t_{il}$ (20)

for all $[k, l] \in Q_i$
for all i

$x_{ik} - x_{il} + My_{kl} \geqslant t_{ik}$ (21)

$x_{jl} - x_{ik} + M(l - y_{kl}) \geqslant t_{jl}$ for all $[k, l] \in N_p$ (22)
for all p

$x_{ik} - x_{jl} + My_{kl} \geqslant t_{ik}$ $i \neq j$ (23)

$$x_{ik} \geqslant \begin{cases} t_{ik} & \text{for } i, k \in I_i, \text{ for all } i \\ 0 & \text{for all other } i, k \end{cases} \qquad (24)$$

$y_{kl} = 0,1$ for all k, l (25)

Constraint (19) imposes that the operations of each part are processed according to the precedences required. Constraints (20) and (21) ensure that any two operations belonging to the same part cannot be processed at the same time. Constraints (22) and (23) ensure that a machine cannot process more than one part at the same time. Constraints (24) and (25) impose nonnegativity and integrality, respectively.

The application of the scheduling model is illustrated in the following example.

Example 7

On two machines, schedule eight operations belonging to three parts shown in Figure 21. The data required are given in Table 2.

Using scheduling model (18)–(25) to solve this problem, the optimal schedule $x_{13} = 26$, $x_{21} = 12$, $x_{22} = 4$, and $x_{33} = 10$ is generated (see Figure 21). The makespan $F_{\max}$ of the schedule in Figure 22 is 26.

The formulation (18)–(25) involves machines and operations only. In order to incorporate other resources such as tools, fixtures, and so on, as well as due dates, the formulation (18)–(25) needs to be extended.

Define

m = number of types of resources
r_s = number of resources of type s, $s = 1, \ldots, m$
$d_{(i)}$ = due date of part P_i
N_q = set of operations using resource p
q = $1, \ldots, r_s$
s = $l, \ldots, m$

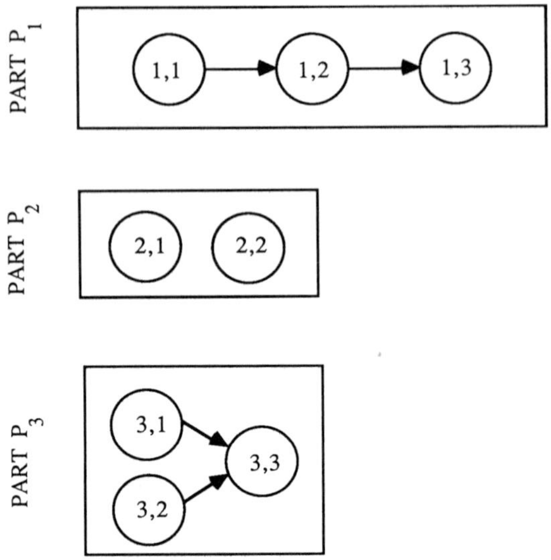

Figure 21. Structure of three parts.

The extended formulation of the *n* operation *m* machine scheduling problem with precedence constraints, limited resources, and due dates is presented below.

$$\min \sum_{i=1}^{n} x_{(i)} \tag{26}$$

$$\text{s. t. } x_{il} - x_{ik} \geqslant t_{il} \qquad \text{for all } [k, l] \in R_i, \text{ for all } i \tag{27}$$

$$x_{il} - x_{ik} + M(l - y_{kl}) \geqslant t_{il} \qquad \text{for all } [k, l] \in Q_i \tag{28}$$

$$x_{ik} - x_{il} + My_{kl} \geqslant t_{il} \qquad \text{for all } i \tag{29}$$

TABLE 2
Scheduling Data

	Part number							
	1			2		3		
Operation number	1	2	3	1	2	1	2	3
Machine number	1	1	2	1	2	2	1	2
Processing time	3	5	6	8	4	9	2	7

Figure 22. The optimal schedule.

$$x_{jl} - x_{ik} + M(l - y_{kl}) \geqslant t_{jl} \qquad \text{for all } [k, l] \in N_q \text{ for all q} \qquad (30)$$

$$x_{ik} - x_{jl} + My_{kl} \geqslant t_{ik} \qquad i \neq j \qquad (31)$$

$$x_{(i)} \leqslant d_{(i)} \qquad \text{for all } i \qquad (32)$$

$$x_{ik} \geqslant \begin{cases} t_{ik} & \text{for } i, k \in I_i, \text{ for all } i \\ 0 & \text{for all other } i, k \end{cases} \qquad (33)$$

$$y_{kl} = 0{,}1 \qquad \text{for all } k, l \qquad (34)$$

Constraint (27) imposes that the operations of each part are processed according to the precedences required. Constraints (28) and (29) ensure that any two operations belonging to the same part cannot be processed at the same time. Constraints (30) and (31) ensure that any resource cannot process more than one part at the same time. Constraint (32) imposes due dates. Constraints (33) and (34) impose nonnegativity and integrality, respectively.

The application of the scheduling model (26)–(34) is illustrated in the following example.

Example 8

On two machines schedule seven operations belonging to three parts shown in Figure 23. The data required are given in Table 3. Solving the scheduling model for the data in Table 3, the optimal schedule is generated (see Figure 24).

The schedule in Figure 24 was generated under the assumption that only one copy of each tool was available. Relaxing this assumption results in the schedule $x_{11} = 2$, $x_{12} = 6$, $x_{22} = 4$, and $x_{33} = 8$ presented in Figure 25. The completion time of part 3 is $C_2 = 8$ rather than $C_1 = 10$ for the solution in Figure 24. As one can see in Figure 25, two identical tools 1 are used for performing operations (1,1) and (3,2).

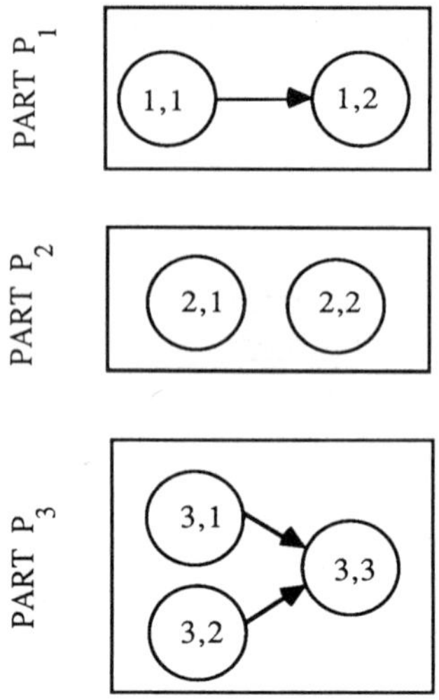

Figure 23. Structure of three parts.

TABLE 3
Scheduling Data

	Part number						
	1		2		3		
Operation number	1	2	1	2	1	2	3
Machine number	1	1	1	2	2	1	2
Processing ime	2	3	1	2	3	2	1
Tool number	1	1	2	2	2	1	2

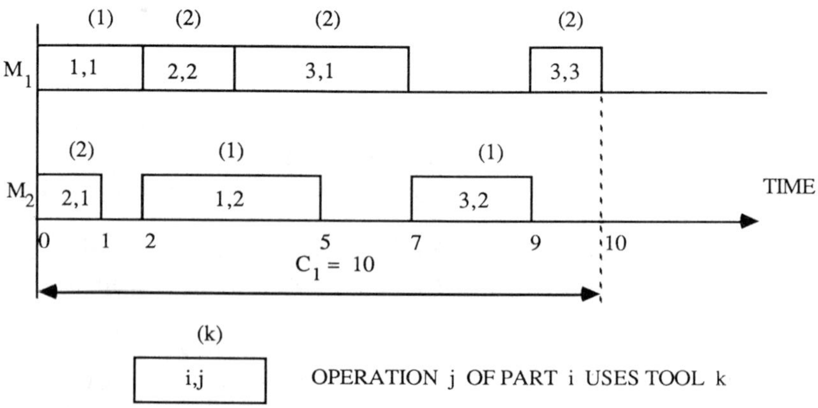

Figure 24. The optimal schedule with limited number of identical tools.

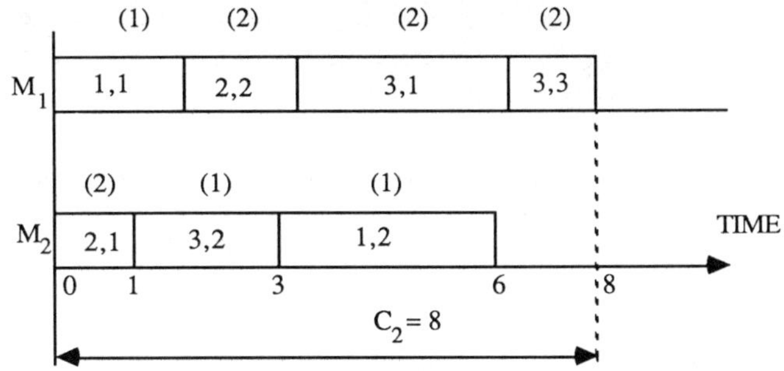

Figure 25. The optimal schedule with unlimited number of identical tools.

VIII. Knowledge-Based Scheduling System

Numerical algorithms have traditionally been used for solving scheduling problems. The approach presented in this section involves not only algorithms but also declarative and procedural knowledge and an inference engine, all implemented as a knowledge-based scheduling system (KBSS). The KBSS performs the following two basic functions:

- Selects an algorithm for the problem considered
- Controls the schedule generation procedure of the algorithm selected

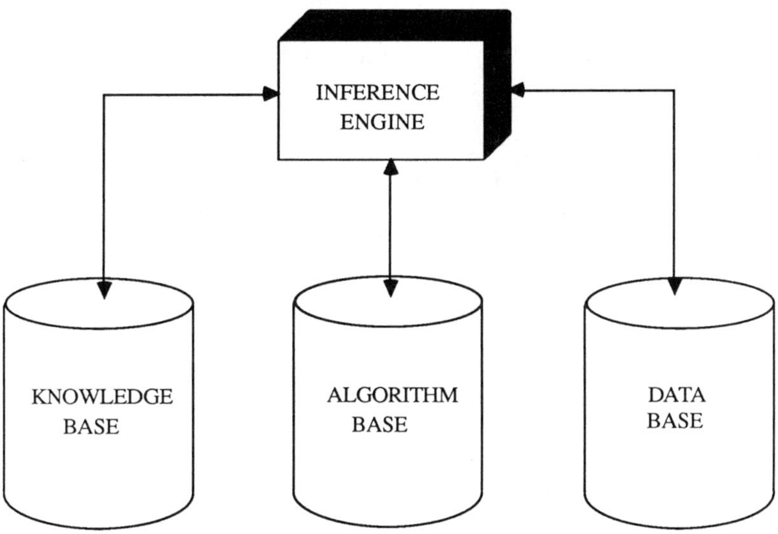

Figure 26. Structure of the knowledge-based scheduling system (KBSS).

The knowledge-based system is built using the tandem architecture proposed in Kusiak (1987). The basic components of the KBSS is shown in Figure 26. (Full details of KBSS and computational results are presented in Kusiak, 1990.)

A. COMPONENTS OF THE KNOWLEDGE-BASED SYSTEM

As shown in Figure 26, the scheduling system KBSS involves

- Knowledge base
- Algorithm base
- Data base
- Inference engine

1. Knowledge Base

Knowledge in KBSS has been acquired from experts as well as the scheduling literature. Frames are used to represent the declarative knowledge related to the description of scheduling problems, parts and operations, and the schedules generated. Four sample frames for a scheduling problem, part, operation, and the schedule generated are presented below.

Frame 1

```
(Problem_number (Problem_type (e.g., flow shop))
  (Problem_features
      (Number_of_parts,
       Number_of_operations,
       Number_of_precedence_constraints)))
```

Frame 2

```
(Part_number (Due_date (value))
             (Number_of_operations (value))
             (Number_of_process_plans (value))
             (Process_plan_0 ( operation_number,
                     processing_time,
                     machine_number,
                     tool_number,
                     pallet_fixture_number,
                     material_handling_carrier_number))
             •
             •
             •
             •
```

```
(Process_plan_p ( operation_number,
                  processing_time,
                  machine_number,
                  tool_number,
                  pallet_fixture_number
                  material_handling_carrier
                  _number)))
```

Frame 3

```
(Operation_number (Belongs to (part_number))
    ((Preceding_op (op_1, op_2, ..., op_a))
     (Successive_op (op_a+1,op_ a+2, . . . ,
     op_b))))
```

Frame 4

```
(Schedule_of_problem (Problem_number)
        (Generated_by (algorithm_number))
        (Idle_time (machine_1, machine_2, ...,
        machine_m))
        (Completion_time (part_, part_2, ..., part
        _n))
        (Average_utilization_rate ( machine,
                                    tool,
                                    pallet_fixture,
                                    material_handling
                                    _carrier))
```

Frame 1 characterizes a scheduling problem considered and is used to select an appropriate scheduling algorithm. Knowledge represented by Frames 2 and 3 is used to generate schedules and select alternative process plans. Frame 4, representing the schedules generated, is used for rescheduling.

The procedural knowledge of the knowledge-based system is in the form of production rules. To handle different problems the production rules are divided into following three classes:

Class 1 selects an appropriate algorithm to solve the problem considered.
Class 2 controls the procedure of selecting alternative process plans and modifying the sequence of the priority rules in the heuristic algorithm.
Class 3 evaluates the schedules obtained and performs rescheduling.

Several sample production rules in each class are presented below.

Class 1

Rule 11

IF the scheduling problem considered is a two-machine flow shop problem

THEN solve it using the Johnson algorithm (Kusiak, 1986).

Rule 12

IF the scheduling problem involves less than 5 resources

AND number of operations is less than 60

AND alternative process plans are not available

THEN use the mixed integer linear programming formulation (see the formulation (18)–(25) presented in this chapter).

Rule 13

IF the scheduling problem involves greater than 5 resources

AND number of operations is greater or equal to 60

AND alternative process plans are available

THEN use the heuristic algorithm (presented in Section IX).

Class 2

Rule 24

IF an alternative process plan is specified for an operation that is not scheduleable due to unavailability of resources listed in the basic process plan

AND the required alternative resources for the operation are available

AND the sum of the waiting and processing time for the operation in the basic process plan is longer than one in the alternative process plan

THEN replace the basis process plan with the corresponding alternative process plan

AND add the corresponding operation to the set of schedulable operations.

Rule 25

IF more than one operation has been added to the set of schedulable operations using production rule 24

THEN select one operation with the alternative processing time closest to the value of the corresponding basic processing time.

Class 3

Rule 36

IF a part in a partial (or final) schedule generated by the heuristic algorithm does not meet the required due date

THEN schedule the part ensuring that the due date is satisfied

AND reschedule other parts using the heuristic algorithm (presented in Section IX).

Rule 37

IF the heuristic algorithm (Section IX) is used for rescheduling

AND the starting time of an operation has been fixed for a machine

AND there is an idle time between the first operation and the rescheduled operation

THEN attempt to reduce the idle time by using production rule 39 (not presented here).

Rule 38

IF all due dates have been met in the final schedule

THEN accept the schedule.

2. Algorithm Base

The algorithms discussed in this chapter as well as other existing algorithms are stored in the algorithm base. Additional algorithms can be easily incorporated into the base. The heuristic algorithm presented later in this section is the most likely to be used for solving scheduling problems in large-scale automated manufacturing systems.

3. Inference Engine

The inference engine in the knowledge-based scheduling system (KBSS) controls the procedure of triggering rules in the knowledge base and the procedure of schedule generation of an algorithm. One of the greatest advantages of the tandem system architecture is the simplicity of the inference engine. The inference engine in KBSS employs a forward chaining control strategy. In a given class of rules it attempts to fire all the rules which are related to the context considered. If a rule is triggered, that is, the conditions are true, then the actions of the rule triggered are carried out. Some rules stop the search of the inference engine and switch the control process to the algorithm.

The inference engine maintains a list of the rules which have been fired. This list is called "explain." The rules in "explain" are placed in the order that they were fired. The list forms a basis for building an explanation facility.

IX. The Heuristic Algorithm

In many manufacturing systems, one associates with each part a basic process plan and one or more alternative process plans. A process plan specifies the operations belonging to the part, processing times of these operations, and the resources required, such as machines, tools, pallet/fixtures, and so on.

Before the heuristic algorithm are presented, the following notation and definitions are introduced:

I = set of all operations
K = set of all parts
IP_k = set of operations belonging to part P_k, $k \in K$
L = set of all resource types
Q_l = set of resources of type l, $l \in L$
d_k = due date of part P_k, $k \in K$
f_i = completion time of operation i, $i \in I$
rt_i = remaining processing time of operation i, $i \in I$
ns_{ik} = number of successive operations of operation i in part P_k, $i \in I$, $k \in K$
nb_{ik} = number of immediate (directly linked by precedence constraints) successive operations of following operation i in part P_k, $i \in I$, $k \in K$
np_{ik} = number of unprocessed operations in part P_k corresponding to operations i, $i \in I$, $k \in K$
t = current scheduling time
r_{lq} = resource q of type l, $q \in Q_l$, $l \in L$

In particular, the following four types of resources are used:

- Machine ($l = 1$)
- Tool ($l = 2$)

- Pallet/fixture ($l = 3$)
- Material handling carrier ($l = 4$)

Resource r_{lq} is *available* if it can be used without any delay, $q \in Q_l$, $l \in L$. The *status* sr_{lq} of such resource equals 1, otherwise $sr_{lq} = 0$.

A *process plan* $PP_k^{(v)}$ of part P_k is a vector of triplets, each containing operation number, processing time, and set of resources to process the operation. It is denoted as follows:

$$PP_k^{(v)} = [(a;\, t_a^{(v)};\, R_a^{(v)}), \ldots, (i;\, t_a^{(v)};\, R_i^{(v)}), \ldots, (b;\, t_b^{(v)};\, R_b^{(v)})]$$

where

$$R_i^{(v)} = (r_{1q}^{(v)}, r_{2q}^{(v)}, \ldots, r_q^{(v)}),\ q \in Q_l,\ l \in L$$

$a, \ldots, i, \ldots, b$ denote operation numbers

$v = 0$ denotes the basic process plan

$v = 1, 2, \ldots$ denotes an alternative process plan

$t_i^{(v)}$ denotes processing time of operation i using process plan v

For simplicity of further considerations two assumptions are made:

1. It is assumed that the number and type of operations in a basic process plan and the corresponding alternative process are identical. This assumption allows one to switch processing from the basic plan to an alternative process plan and reverse after the operation under consideration has been completed.
2. Without loss of generality, it is assumed that $t_i^{(0)} \leq t_i^{(v)}$, $i \in I$, which holds in practice. A process plan $PP_k^{(v)}$ for part P_k and the corresponding operations is *available*, if each element in $PP_k^{(v)}$ has been specified.

Operation i is *schedulable* at time t, if

1. No other operation that belongs to the same part is being processed at time t
2. All operations preceding operation i have been completed before time t
3. All resources required by the basic process plan to process operation i are available at time t

Based on the preceding definitions, further notation is introduced.

$$\textit{operation status}\quad s_i = \begin{cases} 0, & \text{operation } i \text{ is nonschedulable} \\ 1, & \text{operation } i \text{ is schedulable} \\ 2, & \text{operation } i \text{ is being processed} \\ 3, & \text{operation } i \text{ has been completed} \\ 4, & \text{operation } i \text{ satisfies the first two conditions in the definition of schedulability} \end{cases}$$

resource status $sr_{lq} = \begin{cases} 1, \text{ resource } r_{lq} \text{ is available} \\ 0, \text{ otherwise} \end{cases}$

S_j = set of operations with $s_i = j$, $j = 0, 1, 2, 3, 4$, $i \in I$

st_k = slack time of part P_k, st_k $(d_k - t - \Sigma\, t_i^{(0)})$ for $i \in IP_k$, $k \in K$ $i \in S_0 \cup S_1$

no_{ik} = number of operations in $S_1 \cap IP_k$, $i \in I$, $k \in K$

In the process of schedule generation, an operation might not be processed according to the basic process plan due to unavailability of the resources specified in the basic process plan. The scheduling heuristic presented in this section exits from step 3 and step 6 and enters the inference engine of the knowledge-based system.

Seven priority scheduling rules have been incorporated into the heuristic algorithm:

Rule P1 selects an operation with the largest number of successive operations.
Rule P2 selects an operation belonging to a part with the minimum number of schedulable operations.
Rule P3 selects an operation with the largest number of immediate successive operations.
Rule P4 selects an operation belonging to a part with the largest number of unprocessed operations.
Rule P5 selects an operation with the shortest processing time.
Rule P6 selects an operation belonging to a part with the shortest slack time.
Rule P7 selects an operation arbitrarily.

The priority rules are used in step 2 of the algorithm in the sequence of P1 to P7. If more than one operation is selected by a rule, the next rules perform further selection. Different sequences of the priority rules have been tested for various problems and the results are reported later in this chapter.

Algorithm 6

Step 0. Set current time $t = 0$ and resource status $sr_{lq} = 1$, $q \in Q_l$, $l \in L$

Step 1. Construct the following two sets:
- Set S_0 of nonschedulable operations ($s_i = 0$)
- Set S_1 of schedulable operations ($s_i = 1$)

Step 2. In the set S_1, select an operation i^* based on the following priority rules:

P1: $ns_{i^*,k} = \max_{i \in S_1}\{ ns_{ik} \}, k \in K,$

P2: $no_{i^*,k} = \min_{i \in S_1} \{ no_{ik} \}, k \in K,$

P3: $nb_{i^*,k} = \max_{i \in S_1} \{ nb_{ik} \}, k \in K,$

P4: $np_{i^*,k} = \max_{i \in S_1} \{ np_{ik} \}, k \in K,$

P5: $t_{i^*} = \min_{i \in S_1} \{ t_i^{(0)} \},$

P6: $st_{k^*} = \min_{k \in K} \{ d_k - t - \sum_{i \in S_0 \cup S_1} t_i^{(0)} \}$

P7: break a tie arbitrarily

Step 3. Set

- Operation status $s_{i^*} = 2$ for operation i^* selected in step 2
- Operation status $s_i = 0$ for all the unprocessed operations of the part corresponding to operation i^*

Delete operation i^* from the set of schedulable operations S_1. If $S_1 \cup S_0 = \phi$, stop; otherwise, set

- Remaining processing time $rt_{i^*} = t_{i^*}^{(0)}$
- Resource status $sr_{lq} = 0$, for $r_{lq} \in R_{i^*}^{(0)}$, $q \in Q_l$, $l \in L$

Update sets S_1 and S_0. If $S_1 \neq \phi$, go to step 2. If $S_1 = \phi$ and no resource is available, go to step 4; otherwise update set S_4, enter the inference engine, and return.

Step 4. Construct the set of operations being processed S_2 and

- Calculate completion time $f_i = rt_i + t$, $i \in S_2$
- Set current time $t = f_{\bar{i}} = \min_{i \in S_2} \{f_i\}$
- Set operation status $s_{\bar{i}} = 3$

- Delete operation $\bar{i}$ from S_2
- Set resource status $sr_{lq} = 1$, $r_{lq} \in R_{\bar{i}}^{(0)}$, $q \in Q_l$, $l \in L$
- Set remaining time $rt_i = f_i - t$, $i \in S_2$

Update sets S_1 and S_0.

Step 5. If $S_1 \cup S_0 = \phi$, stop; otherwise, go to step 6.

Step 6. If the set of schedulable operations $S_1 \neq \phi$, go to step 2. If $S_1 = \phi$ and no resource is available, go to step 4; otherwise, update set S_4, enter the inference engine, and return.

The algorithm presented solves scheduling problems with due dates. If the due date is not imposed for a part P_k, the corresponding value of d_k is set as an arbitrary large number.

The preceding algorithm is embedded into the knowledge-based system discussed in the previous section. In the next section, the heuristic algorithm and sample production rules are illustrated with the following numerical example.

Example 9

Schedule 12 operations shown in Figure 27 on three machines. It is assumed that

- Three different tools are available to process the operations
- All other resources are unlimited
- Due dates are not imposed

The following notation is used for resources:

r_{1q} denotes machine (resource type 1), $q = 1,2,3$
r_{2q} denotes tool (resource type 2), $q = 1,2,3$

The machine and tool status are represented as follows:

sr_{1q} denotes machine status, $q = 1,2,3$
sr_{2q} denotes tool status, $q = 1,2,3$
t_i denotes the processing time $t_i^{(0)}$ of operation i in the basic process plan

The basic process plans of the four parts are shown below:

$PP_1^{(0)}$: [(1; 4; 2, 2),(2; 5; 1, 3),(3, 2; 3, 2)]

$PP_2^{(0)}$: [(4; 6; 1, 3),(5; 3; 2, 2),(6; 3; 3, 1)]

$PP_3^{(0)}$: [(7; 3; 3, 1),(8; 3; 1, 2),(9; 6; 3, 1),(10; 2; 1, 3)]

$PP_4^{(0)}$: [(11; 4; 3, 2),(12; 3; 2, 3)]

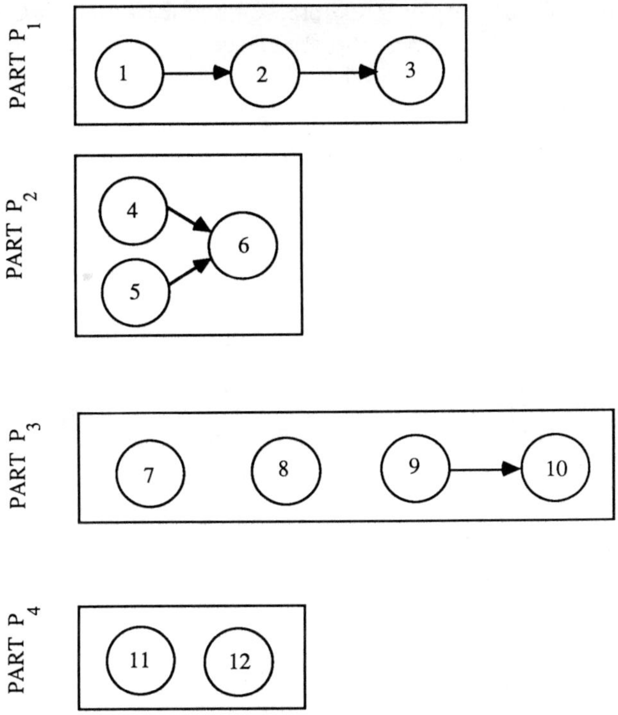

Figure 27. Parts with operations and precedence constraints.

The alternative process plans for the four parts are

$PP_1^{(1)}$: [(1; 6; 3, 12), (2; 6; 2, 2), (3; 4; 1, 1)]

$PP_1^{(2)}$: [(1; 7; 1, 3), (2; 7; 1, 2), (3; 5; 1, 3)]

$PP_2^{(1)}$: [(4; 6; 2, 2), (5; 4; 3, 1), (6; 5; 1, 2)]

$PP_2^{(2)}$: [(4; 8; 3, 1), (5; 8; 1, 3), (6; 5; 2, 3)]

$PP_3^{(1)}$: [(7; 4; 3, 2), (8; 5; 3, 3), (9; 7; 2, 1), (10; 2; 3, 2)]

$PP_3^{(2)}$: [(7; 4; 2, 2), (8; 5; 2, 1), (9; 9; 1, 3), (10; 4; 1, 2)]

$PP_4^{(1)}$: [(11; 4; 1, 3), (12; 5; 1, 2)]

$PP_4^{(2)}$: [(11; 4; 3, 1), (12; 6; 3, 3)]

Note that, for any triplet in the preceding process plans, the first element denotes operation number, the second denotes processing time, and the third pair denotes the required machine number (resource type 1) and tool number (resource type 2).

Solution Procedure

Step 0. Set current time $t = 0$ and $sr_{lq} = 1$, $l = 1, 2$, $q = 1, 2, 3$.

Step 1. Construct the following two sets:
- $S_0 = \{2, 3, 6, 10\}$
- $S_1 = \{1, 4, 5, 7, 8, 9, 11, 12\}$

Step 2. Using priority rule P1, operation 1 is selected.

Step 3. Set
- $s_1 = 2$
- $s_2 = 0$, $s_3 = 0$

Since $S_0 \cup S_1 = \{2, 3, 4, 5, 6, 7, 8, 9, 10, 11, 12\} \neq \phi$, set
- $rt_1 = t_1 = 4$
- $sr_{12} = 0$, $sr_{22} = 0$

Set of schedulable operations $S_1 = \{4,7,9\} \neq \phi$. Go to step 2.

Step 2. Using priority rules P1 and P2, operation 4 is selected.

Step 3. Set
- $s_4 = 2$
- $s_5 = 0$, $s_6 = 0$

Since $S_0 \cup S_1 = \{2, 3, 5, 6, 7, 8, 9, 10, 11, 12\} \neq \phi$, set
- $rt_4 = t_4 = 6$
- $sr_{11} = 0$, $sr_{23} = 0$

Set of schedulable operations $S_1 = \{7, 9\} \neq \phi$. Go to step 2.

Step 2. Using priority rule P1, operation 9 is selected.

Step 3. Set operation statutes
- $s_9 = 2$
- $s_7 = 0$, $s_8 = 0$, $s_{10} = 0$

Since $S_0 \cup S_1 = \{2, 3, 5, 6, 7, 8, 10, 11, 12\} \neq \phi$, set
- $rt_9 = t_9 = 6$
- $sr_{13} = 0$, $sr_{21} = 0$

Set of schedulable operations $S_1 = \phi$. Since $S_1 = \phi$ and no resource is available, go to step 4.

Step 4. Construct the set of nonschedulable operations $S_2 = \{1, 4, 9\}$ and
- Calculate completion time $f_1 = 4, f_4 = 6, f_9 = 6$.
- Set current time $t = f_1 = \min\{4, 6, 6\} = 4$
- Set $s_1 = 3$
- Delete operation 1 from S_2
- Set $sr_{12} = 1, sr_{22} = 1$
- Set remaining time $rt_4 = 6 - 4 = 2, rt_9 = 6 - 4 = 2$

Set of schedulable operations $S_1 = \phi$ (there is no operation with the status equal to 1).

Step 5. Since $S_1 \cup S_0 = \{2, 3, 5, 6, 7, 8, 10, 11, 12\} \neq \phi$, go to step 6.

Step 6. Since $S_1 = \phi$ and machine 2 and tool 2 are available, set $S_4 = \{2, 11, 12\}$ and enter the inference engine. The inference engine activates production rule 24 in Class 2 to select an alternative process plan for operation 2.

Rule 24:

SINCE the alternative process plan is specified for operation 2, in S_4

AND the alternative resources for operation 2 (r_{12} and r_{22}) are available

AND the sum of the waiting and processing time for operation 2 in the basic process plan ($2 + 5 = 7$) is longer than one in alternative process plan (6)

THEN operation 2 is moved from S_4 to S_1

AND the basic process plan of operation 2 is replaced with the corresponding alternative process plan (2,6,2,2)

After the basic process plan has been replaced with the alternative process plan, the inference engine transfers the information to the heuristic algorithm and operation 2 is scheduled by the algorithm.

The Gantt chart of the final schedule obtained after seven iterations is shown in Figure 28. Assume that, in the final schedule, due date $d_4 = 12$ for part P_4 (operation 12) is violated. The appropriate production rules are activated and part P_4 as well as other parts are rescheduled. The new schedule generated is shown in Figure 29. In this schedule, the due date for part P_4 is satisfied.

Note that, in the preceding problem, only machines and tools have been

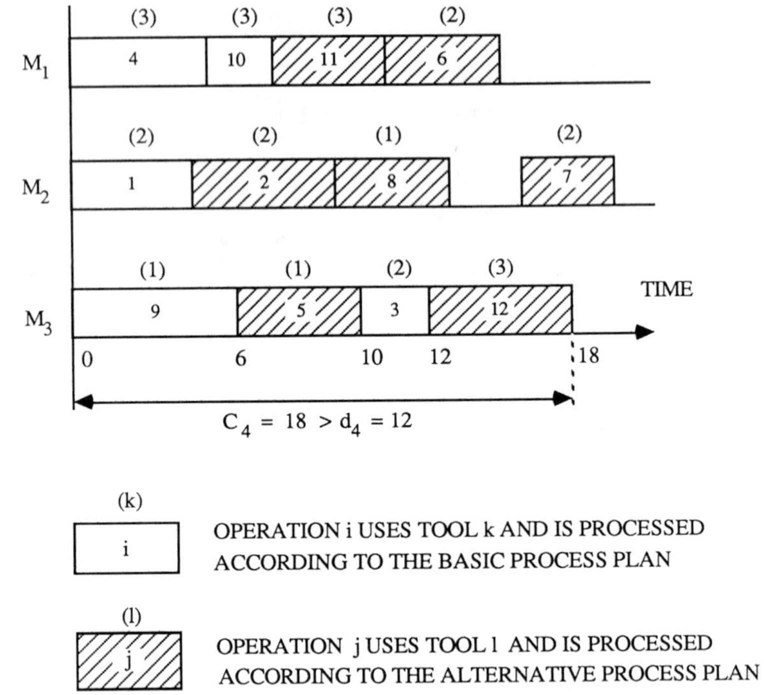

Figure 28. The final schedule.

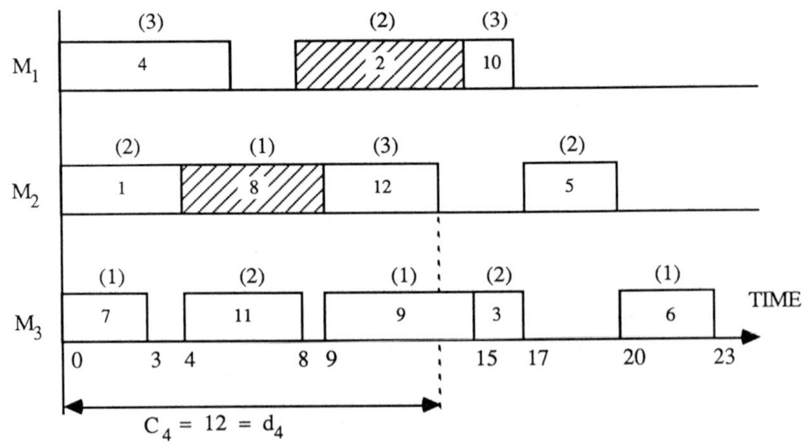

Figure 29. The modified schedule of Figure 28.

considered. The KBSS is general enough to consider other resources such as pallets/fixtures, material handling carriers, and so on.

Appendix: Proof of Theorem 1

Without loss of generality, consider the two-level simple digraph G_s in Figure 30. Due to the precedence constraints between subassembly A_1 and parts $P_a, \ldots,$ P_d, and those between subassembly A_2 and parts $P_l, \ldots, P_n$, scheduling nodes of the digraph according to the MLDF rule results in the following makespan:

$$C_{\max} = \sum_{j=a}^{d} t(P_j) + \sum_{i=l}^{n} t(P_i) + t(A_2)$$

$$+ \max\left\{ t(A_1) - \sum_{i=l}^{n} t(P_i), 0\right\}$$

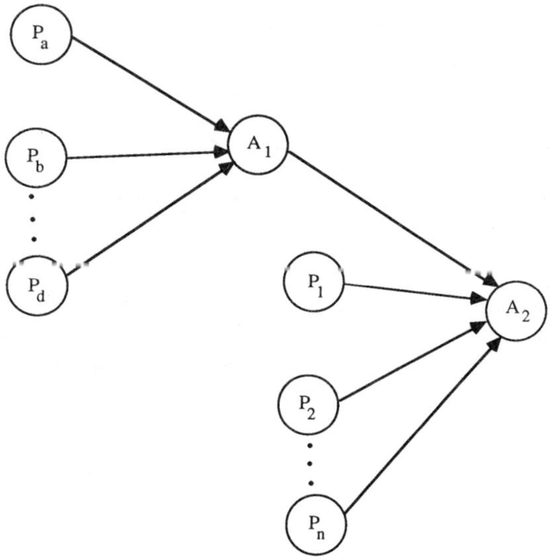

Figure 30. Simple digraph G_s.

where $t(*)$ is the processing time of a part or subassembly*.

If any P_k, $k = 1, 2, \ldots, n$ is scheduled before P_j, $j = a, \ldots, d$, the makespan can be expressed as follows:

$$C^k_{\max} = \sum_{j=a}^{d} t\,(P_j) + \sum_{i=l}^{n} t\,(P_i) + t\,(A_2)$$

$$+ \max \{ t\,(A_1) - \sum_{\substack{i=l \\ i \neq k}}^{n} t\,(P_i), 0\}$$

It is clear that $C_{\max} \leqslant C^k_{\max}$, $k = 1, 2, \ldots, n$

References

Baker, K. R. (1974). "Introduction to Sequencing and Scheduling." Wiley, New York.

Finke, G., Claus, A., and Gunn, E. (1984). A two-commodity network flow approach to the traveling salesman problem. *Congressus Numerantium,* **41,** 167–178.

French, S. (1982). "Sequencing and Scheduling: An Introduction to the Mathematics of the Job-Shop." Wiley, New York.

Johnson, S. M. (1954). Optimal two- and three-stage production schedules with set-up times included. *Nav. Res. Logistics Q.* **1,** 61–68.

Kurisu, T. (1976). Two-machine scheduling under required precedence among jobs. *J. Oper. Res. Soc. Jpn.* **19**(1), 1–13.

Kusiak, A. (1986). Efficient implementation of Johnson's scheduling algorithm. *IIE Trans.* **18**(2), 215–216.

Kusiak, A. (1987). Artificial intelligence and operations research in flexible manufacturing systems. *Inf. Process. Oper. Res (INFOR)* **25**(1), 2–12.

Kusiak, A. (1988). Scheduling flexible machining and assembly systems. *Ann. Oper. Res.* **15,** 337–352.

Kusiak, A. (1990). "Intelligent Manufacturing Systems." Prentice-Hall, Englewood Cliffs, New Jersey.

Kusiak, A., and Finke, G. (1987). Modeling and solving the flexible forging module scheduling problem. *Eng. Optim.* **12**(1), 1–12.

Lawler, E. L., Lenstra, J. K., Rinnooy Kan, A. H. G., and Shmoys, D. B. (1985); "The Traveling Salesman Problem: A Guided Tour of Combinational Optimization." Wiley, New York.

Martello, S. (19830. An enumerative algorithm for finding Hamiltonian circuits in a directed graph. *ACM Trans. Math. Software* **14,** 256–268.

Schrage, L. (1984). "Linear, Integer and Quadratic Programming with LINDO." Scientific Press, Palo Alto, California.

CHAPTER 2

Computer Control in Flexible Manufacturing Systems

Vinay S. Sohoni
Strategic Program Advisor
IBM Corporation
Somers, New York

I. Introduction

In today's environment, even large industries are forced to introduce new and improved products more frequently to maintain or increase their share of the world market. The increasing pace of technological development and the fierce competition require industries to introduce high-quality products at competitive prices. Hard automation and inflexible manufacturing facilities can result in delays and higher costs for introduction of new products. However, manufacturing facilities with flexible automation can react to changing market environments by producing products with the desired quality and costs. Computer control of machines and material handling systems makes a manufacturing facility more flexible since simple software changes can produce a variety of different parts from the same manufacturing facility.

It is difficult to describe a "standard" flexible manufacturing system (FMS) control environment as the demands on such an environment depend on the varied requirements of different manufacturing applications. Many articles and papers have appeared in the literature detailing only a specific aspect of the FMS computer control. Rembold *et al.* (1985) described computer control of the individual manufacturing equipment as well as the control of parts handling systems. Young (1981) outlined the software strategies to implement the information flow and the control requirements of the FMS. Talvage and Hannam (1988) documented their experiences with various manufacturing firms in various stages of automation along with the simulation techniques in designing FMSs. Gershwin *et al.* (1986) presented their interpretation of the recent progress in manufacturing systems decomposed through various hierarchies from the perspective of control. The Economic Commission for Europe (1987) reviewed the present state and the current trends in software for industrial automation.

In this chapter an attempt is made to provide the reader a consistent view of computer control in the FMS environment. Section II describes a generic product manufacturing line consisting of work centers, subassembly lines, and the main assembly line. This generic manufacturing line is used to describe the hierarchy of computer control in FMSs. This hierarchical structure and its advantages/disadvantages are described in Section III. Sections IV, V, and VI describe the computer control functions in work centers, subassembly/main assembly lines, and the overall computer supervision of FMSs, respectively.

II. Flexible Manufacturing System Composition— A Generic Scenario

Figure 1 is a schematic of a generic discrete manufacturing facility for a product. It contains a main assembly line and a series of subassembly lines, feeding subassemblies at the correct locations within the main manufacturing process. Each assembly line is made up of several work centers. The number of work centers in an assembly line is a function of the number of discrete manufacturing steps required to produce the assembly. Each work center consists of machines and robots. The types of machines installed in a work center are dictated by the manufacturing processes performed in it. The movement of the work product within the work center is shown via robots in Figure 1. The work centers are shown to be logistically interconnected via conveyor belts. Interwork center logistics can also be accomplished by robots, automated guided vehicles (AGVs), and so on.

In a fully automated manufacturing line, every element is computer-controlled. Controlling the isolated automated machine or the hard automated section of a line is much simpler than control of the entire automated line. It requires a tightly coupled information system with the manufacturing environment. The

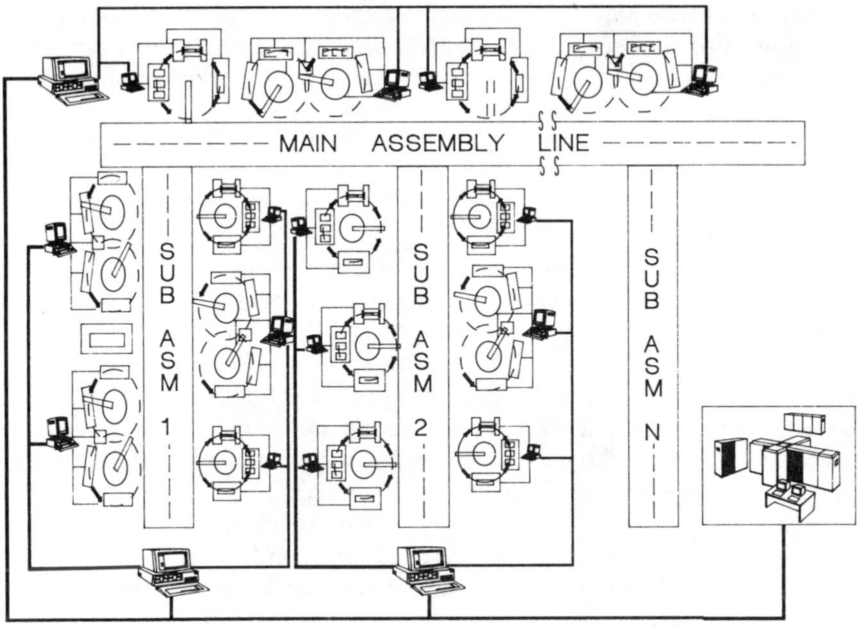

Figure 1. Generic manufacturing facility.

information system consists of such elements as order entry systems, quality systems, maintenance and engineering systems, and so on. In the absence of human intervention, the computer-controlled manufacturing facility also requires a periodic statistical check at several logical points within the process with predefined statistical norms to ensure the quality of the resultant product.

Practically speaking, even in a fully automated line, as shown in Figure 1, a handful of operators/engineers are required to troubleshoot the machines and maintain the flow of production. Especially in a fully automated line, it is also important to have the design of the product, the quality of incoming material, the design of the equipment, and systems such that the product can be produced with zero defects. A well-designed automated line avoids automation of waste (unnecessary motion, processing, product movement, etc.). Also, every member of the automated line must produce product at a consistent, common beat (*takt* time). *Takt* is a German word meaning the tempo or beat in music. *Takt* time is used in just-in-time or continuous flow manufacturing to coordinate the total production at a consistent pace. This beat determines the rate of production. To produce a variety of end products, the product design must be modular. Each module is produced at a work center or subassembly line. Ideally, the change in software should be able to alter the functionality and/or the composition of the module/subassembly. Figure 1 also shows the hierarchy of the computer control in an FMS. Depending upon the product design, FMSs may range anywhere from a single work center to the fully automated facility shown in Figure 1.

III. The Hierarchy of Computer Control

One of the early applications of a digital computer in an industrial facility was for plant monitoring and supervisory control, as documented by Garrett and McHenry (1981). Figure 2 shows an evolution of the use of digital computers in industrial processes.

As illustrated in Figure 2, Dupont-Gatelmand (1981) documented that the next step in evolution after the supervisory control is the computer numerical control (CNC) of the machines followed by direct digital control (DDC) for a group of multiple numerical control (NC) machines.

In the DDC, as written by Lukas (1986), a computer reads and directly processes measurements, calculates the proper control outputs, and sends the control commands to the activation devices. In the initial implementations of DDC, backup analog control systems were used to avoid the ill effects of computer failures. In spite of the early computer hardware reliability problems, DDC demonstrated many advantages over analog control systems. They included the use of complex logic to calculate more accurately the control command values, ease of data logging, data trending, alarming, and so on. It also avoided the common problem of set-point drifts associated with analog devices. Several different system architectures evolved for the DDC systems in the late 1970s.

However, a central computer was a dominant feature of all the variations of these architectures. The single largest disadvantage of these architectures, as pointed out before, was the single-point failure of the central computer, which could shut down the process. It necessitated the expense of a second computer as backup. Another disadvantage was that the software for the central computer was very complex and required a team of software/hardware experts to change and maintain the software. It also had limited expansion capability and, when the expansions were made, they were very expensive.

In the mid 1960s the distributed control system architecture was brought forward as a viable option. The technology to implement the DDC in a cost-effective manner was not available until the early 1970s. The price of computers decreased significantly and personal computers could be used economically on the factory floor. A number of production lines with distributed control systems have begun to emerge since the late 1970s.

As shown in Figure 1, every work center has a dedicated control computer. The work center computer is responsible for making production happen in its

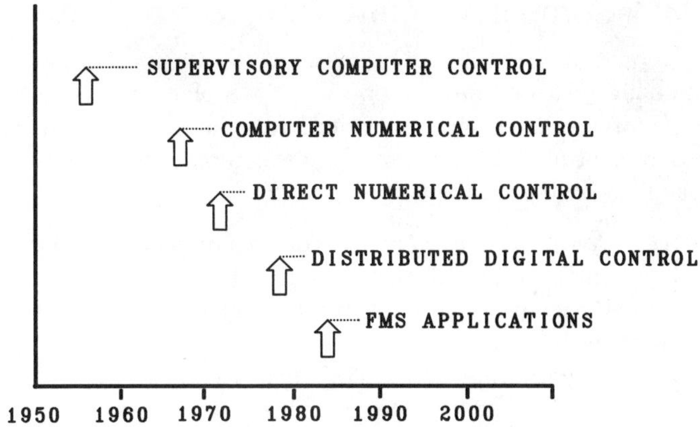

Figure 2. Evolution of the use of digital computers.

work center. It communicates with each tool and robot and downloads the programs, resulting in appropriate commands for all equipment. It is also responsible for ascertaining the health of all the components in the work center. Details of control functions executed by the work-center computer are described in Section IV.

The next level in the hierarchy of computer control is the subassembly/main assembly line computer. In general, this computer coordinates and controls the manufacturing activities within a section of the manufacturing facility. This computer communicates to each work-center computer the type of product to be made and all the appropriate commands. It also serves as a backup to any one of the failed work-center computers in its jurisdiction. The details of the control functions to be executed by the subassembly/main assembly line computer are described in Section V.

As shown in Figure 1, the supervisory computer is at the highest level in the hierarchy of computer control. This computer does overall production planning and scheduling and communicates with the subassembly/main assembly line computers (Koren, 1983). In the event of the failure of any of the subassembly/main assembly computers, the supervisory computer takes over the tasks of the failed computer. Detailed descriptions of the control functions of the supervisory computer are given in Section VI.

The key for producing economically different products or part numbers from the automated line shown in Figure 1 is the flexibility provided by the computer control. Simple software changes can dictate the automated manufacturing line to produce different part numbers by changing the number of subassemblies, the manufacturing process in designated work centers, and the logistics flow of the parts.

IV. Computer Control in a Work Center

The task of controlling total production in an FMS plant is the management of a complex set of machines interconnected with the automated parts transfer mechanisms. To control a real-time process, the computer must do the following, as documented by Williams (1988):

1. *Process Control Commands:* The control computer must have the software capability to direct the hardware devices to do their tasks. The hardware includes the actual machines that perform the manufacturing operations and the logistics mechanisms.
2. *Process-Initiated Interrupts:* The control computer must receive and respond to the signals received from the process. Depending upon the importance of the signal, the computer may have to abort its current operation and perform a priority task.
3. *Periodic Time-Initiated Events:* The control computer must periodically collect management and status information. This is important for creating a history base and performing trend analyses.
4. *System and Program-Initiated Events:* The work-center computer is connected to subassembly/main assembly control computer. Therefore, it must handle communication and data transfer with the higher level computers.

The lowest level control computer in the hierarchy of FMS control is the work-center control computer. It controls a small collection of hardware, as shown in Figure 1. The type and size of the work-center control computer is a function of the number of machines in the work center, the complexity of the manufacturing process, and the required response time, as pointed out by Sheff (1986). Figure 3 shows a set of control functions the work-center control computer must perform. A brief description of each of these control functions is given in the remainder of this section.

1. *Schedule Execution:* The work-center computer must execute the work-center schedule. It must marshal all the resources in the work center to complete the schedule activities. It should also communicate the execution progress to the subassembly/main assembly computer.
2. *Process Device Control:* The control computer can exercise true device control functions in the open-loop or closed-loop mode. The preplanned control is of the open-loop mode in which a predetermined standard set of commands or a sequence of commands with intermediate logical checks are executed. In a closed-loop model, the control computer attains a desired value of a control variable by comparing its actual value with the desired value and uses the error signal as a control input. All devices which do not have their own control computer are controlled by the work-center control computer in this manner.

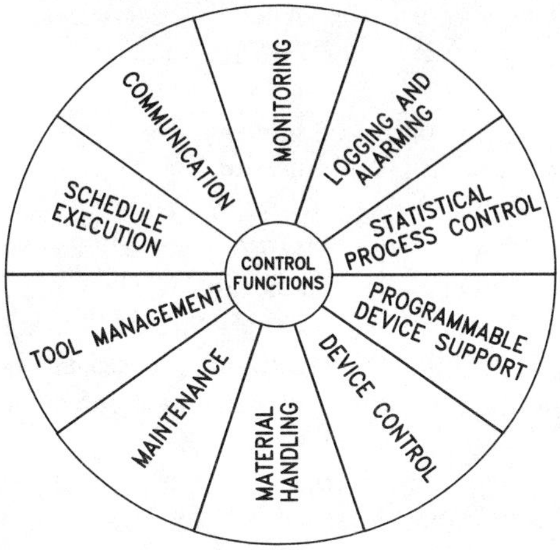

Figure 3. Work-center computer control functions.

3. *Programmable Device Support:* The control computer is responsible for downloading the part programs to the devices in the work center that have their own control computers. The downloading is coordinate to the work-center production schedule.

4. *Tool Management:* This control function includes management of all reusable resources such as drills, bits, gauges, and so on, with the work center. It should keep the inventory of all tools and monitor tool wear.

5. *Maintenance:* The work-center control computer must keep track of the health of each piece of equipment within the work center. This includes keeping a maintenance history and creating preventive maintenance schedules. The maintenance function is directly related to the availability of the work center for production.

6. *Material Handling:* The control computer is responsible for the flow of material into the work center, movement of material within it, and the flow of material exiting it. As shown in Figure 1, the devices used for material handling are the conveyors, robots, AGVs, and so on.

7. *Statistical Process Control:* In a completely automated FMS plant, it is necessary to perform a statistical check within the work center to a predefined statistical norm. The control computer must perform such checks periodically and take appropriate actions to correct any variations in the manufacturing process within the work center.

8. *Communication:* The control computer must communicate with every device within the work center. It must also communicate with the sub-assembly/main assembly computer. Orderly and accurate communication is

needed to avoid costly waste. The exact product schedule will be communicated to the work-center computer by the subassembly/main assembly computer.

9. *Monitoring:* The control computer serves as an alert and indefatigable supervisor. The most important monitoring task is that of process monitoring. This usually requires that the control computer establishes the status of the instruments and the process variables, the status of the equipment within the work center, and the status of the product itself. The control computer can also monitor indirect measurements, as shown by Savas (1985). These measurements are a function of several directly monitored process attributes. The computer supervision is an important function in producing a defect-free product in a work center.

10. *Data Logging and Alarming:* The work-center control computer must collect and store the information about all the devices in the work center and the significant events that take place in it. These data will be used to create preventive maintenance schedules and keep the processing capabilities of the devices current. The data can be uploaded to the subassembly/main assembly computer for trend analysis. The data can be used for the alarm management function.

V. Computer Control in Subassembly/ Main Assembly Lines

The control computer in a manufacturing line communicates with the work-center control computers and ensures scheduled production through its line. The subassembly/main assembly line control computer receives its production schedule from the FMS supervisory computer. The primary functions of the subassembly/main assembly line control computers are as follows.

1. *Monitor Production Performance:* The subassembly/main assembly line control computer must monitor the production performance of each work-center control computer in its jurisdiction. If the work-center control computer fails, the subassembly/main assembly line control computer has the responsibility to function as a backup control computer. Therefore, it must be capable of performing the control functions of the work-center control computer given in Section IV.

2. *Database Management:* The subassembly/main assembly line control computer receives and stores all the process and device-related data in its database, as pointed out by (Meister, 1987). It manipulates these data and uploads them to the FMS supervisory control computer for generating production plans.

3. *Production Scheduling:* The subassembly/main assembly line control computer receives primary production and two backup production schedules each day from the FMS supervisory control computer. This process is described in Section VI. The subassembly/main assembly control computer utilizes the backup production schedule in the anomalous operating conditions. It communicates such operating conditions to work-center computers as well as to the FMS supervisory control computer.

4. *Alarm Management Section:* The subassembly/main assembly line control computer is responsible for the alarm management system. The control computer must keep log of the reasons for all the alarms in all the work centers in its jurisdiction.

VI. Flexible Manufacturing System Supervisory Computer Control

As the name implies, the supervisory computer supervises and controls production throughout the entire manufacturing facility. This is the highest and most important level in the hierarchy of computer control. At this level, the demands for various parts are analyzed. Knowing the process involved in the manufacture of each product and the capacity of each of the equipment, a production schedule is created. Simulations are run to understand the capacity pinch-point operations. Akella *et al.* (1984) described a hierarchical production scheduling policy that minimizes the disruptive effects of such disturbances as machine failures. A cost analysis of the schedule is also performed to quantify the unused capacity within each subassembly area. Since an FMS facility is very costly, it is important to utilize the equipment as much as possible, as written by Morton and Smut (1986).

Once the schedule is finalized, material requirements are created. Also, detailed work plans for each subassembly/main assembly work station are developed for each day. Smith *et al.* (1986) made a survey of the scheduling criteria used by FMS users and their relation to the scheduling techniques published in the literature. In general, a rolling one-week firm schedule with the production outlook for the next four weeks appears to be an adequate scheduling scenario for a complex product. That is, at the end of each day a new day is added to the firm schedule and to the outlook. The detailed work plan for each day is downloaded to each subassembly/main assembly computer at the beginning of each day. An overview of the important control functions of the supervisory computer are presented in the remainder of this section.

1. *Production Planning:* Production planning necessarily begins with the analysis of firm demands for each product and the forecast of future sales. The supervisory computer database necessarily contains the description of the production process for each product and the tool capacity in each subassembly/main assembly line. A preliminary production plan for the facility is created for the rolling day that will accommodate the firm and projected demands.

2. *Simulation:* The demands for each product within the production plan should be simulated to understand the pinch-point resources and the intricate subtleties for the product changeover requirements. "What if" scenarios should also be simulated to create alternative production plans. They should include tool failures, part shortages, or quality problems. The data used should be based upon the historical data stored in the supervisory computer. The simulation effort may be very complex since a product may be able to take alternative routes made possible by the FMS. Buzacolt (1983) described a basic philosophy to develop models to estimate performance and key technical issues of FMS. If the preliminary production plan does not appear feasible, the production planning step described above may have to be repeated.

3. *Master Production Schedule:* A detailed production schedule for each day within the rolling week is created for the preliminary production plan, as well as two backup production plans using the "what if" analysis. The details of the master production schedule include the sampling/verification plan for each subassembly and for the final assembly for quality. Fox (1982) gave an overview of the FMS software control functions for variable missions and scheduling procedures.

The master production schedule must be accurate. It will be downloaded each day to the subassembly/main assembly computers and it dictates the total production operations within the FMS.

4. *Capacity Analysis:* The supervisory computer must determine the unused capacity in each subassembly area. This unused capacity may be utilized to produce spare parts such as field replaceable units (FRUs). Also, the supervisory computer must schedule preventive maintenance consistent with the utilization of the equipment. Kumar and Vannelli (1986) presented a flexible decision process to help balance the capacity of work centers.

5. *Communication:* A supervisory computer requires constant communication with the subassembly/main assembly computers to be aware of the status of the equipment as well as the production. It can then compare the status with the simulation results to predict the problem and sound appropriate alarms. In a completely automated FMS, the role of the supervisory computer is very important. It can help to fully utilize FMS hardware and therefore help to manufacture cost-competitive products.

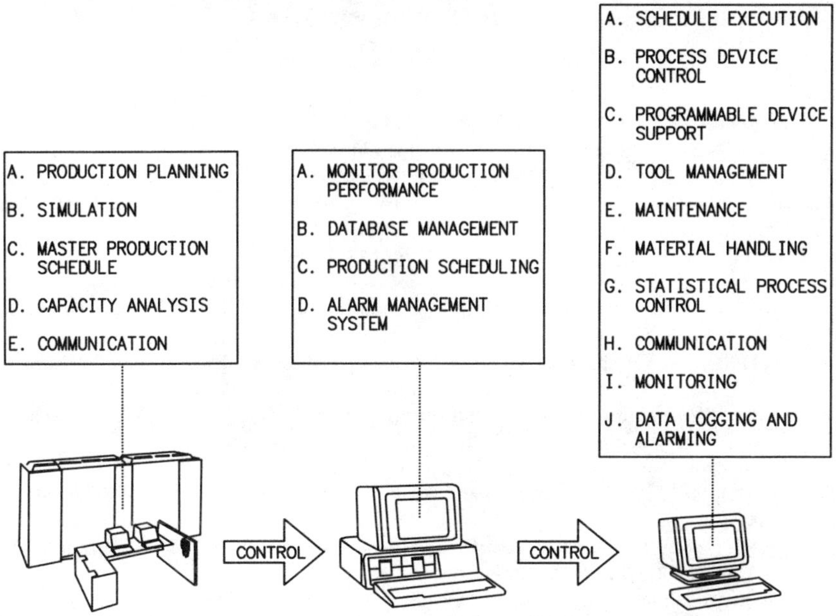

Figure 4. Control functions summary.

VII. Concluding Remarks

In the FMS, a coordinated participation of all computers is needed to control and economically manufacture products. A description of all encompassing computer control functions for a variety of flexible manufacturing systems producing different products is very difficult. A systematic description of the control functions using a generic FMS concept has been given in this chapter. Figure 4 summarizes all control functions performed at each hierarchical level. In recent years, the cost of computers has decreased and computers for process control are more prevalent on the manufacturing floor. However, the hardware/software needed to put the entire FMS under computer control is not presently available. But the rapid advancement in software/hardware technology will make total computer control of FMSs a reality in the near future.

References

Akella, R., Choong, Y., and Gershwin, S. (1984). *IEEE Trans. Components, Hybrids, Manuf. Technol.* **CHMT-7**(3), 215–217.

Buzacolt, J. (1983). *Fall Ind. Eng. Conf. Proc.* pp. 130–137.

Dupont-Gatelmand, C. (1981). *J. Manuf. Syst.* **1**(1), 1–16.

Economic Commission for Europe (1987). *In* "Software for Industrial Automation," pp. 15–68. United Nations, New York.

Fox K. (1982). *Soc. Manuf. Eng.—Autofact 4 Conf.* MS82-419, pp. 1–9.

Garrett, L. T., and McHenry, J. M. (1981). *Hydrocarbon Process.* **60**(12), 103–108.

Gershwin, S. D., Hilderbrant, R. R., Suri, R., and Mitter, S. K. (1986). *IEEE Control Syst. Maga.* **6,** 3–15.

Koren, Y. (1983). *In* "Computer Control of Manufacturing Systems," pp. 249–267. McGraw-Hill, New York.

Kumar, K., and Vannelli, A. (1986). *In* "Proceedings of the Second ORSA/TIMS Conference on Flexible Manufacturing Systems" (K. Stecke and R. Suri, eds.), pp. 204–219. Elsevier, Amsterdam.

Lukas, M. P. (1986). *In* "Distributed Control Systems," pp. 6–17. Van Nostrand-Reinhold, New York.

Meister, A. E. (1987). *In* "Computer Integrated Manufacturing Handbook" (E. Teicholz and J. Orr, eds.), pp. 2-123–2-139. McGraw-Hill, New York.

Morton, T. E., and Smut, T. L. (1986). *In* "Flexible Manufacturing Systems: Methods and Studies" (A. Kusiak, ed.), pp. 151–164. Elsevier, New York.

Rembold, V., Blume, C., and Billmann, R. (1985). *In* "Computer-Integrated Manufacturing Technology and Systems," pp. 387–410 and 500–511. Dekker, New York.

Savas, E. S. (1985). *In* "Computer Control of Industrial Processes," pp. 10–33. McGraw-Hill, New York.

Sheff, B. (1986). *Soc. Manuf. Eng.—Flexible Manuf. Cell Semin.* pp. 1–13.

Smith, M., Ramesh, R., Dudek, R., and Blair, E. (1986). *In* "Proceedings of the Second ORSA/TIMS Conference on Flexible Manufacturing Systems" (K. Stecke and R. Suri, eds.), pp. 478–486. Elsevier, Amsterdam.

Talvage, J., and Hannam, R. G. (1988). *In* "Flexible Manufacturing Systems in Practice," pp. 3–11. Dekker, New York.

Williams, D. J. (1988). *In* "Manufacturing Systems," pp. 152–153. Open University Press, Milton Keynes, England.

Young, R. E. (1981). *Ind. Eng. J.* November Issue, pp. 1–10.

CHAPTER 3

Computer Software for Flexible Manufacturing Systems

Gary W. Fischer
Department of Industrial Engineering
The University of Iowa
Iowa City, Iowa

I. Introduction

Flexibility in manufacturing systems provides an opportunity to capitalize on basic strengths of a company to ensure long-term leadership in manufacturing cost, product, and service to the customer. The degree and type of flexibility specified for a system must be carefully determined to make the best use of capital and to achieve the needed level of productivity. The flexibility of a flexible manufacturing system (FMS) is characterized by how well it responds to change in product designs and to change in production schedules (e.g., part quantities, delivery times, and model mix). For the most part, the degree to which an FMS can respond to either type of change is determined by the design of the system.

Flexible manufacturing systems exist in a variety of configurations, degrees of complexity, and a wide range of sizes. Individual parts produced by an FMS

HANDBOOK OF FLEXIBLE MANUFACTURING SYSTEMS

range in size from small precision components used in instrumentation to very large structural components of construction equipment. The system components found in an FMS vary as much as the parts produced. The significance of the decision to purchase an FMS and the large number of variables to consider suggest the need for computer software to assist in the decision process. The need for computer software extends beyond the purchase to assist the FMS manager in getting the greatest possible utilization of the investment.

Following a brief review of the definition, system concepts, system components, and typical functions performed by FMS software, some of the system issues and considerations for selecting FMS software will be examined. Many of the conceptual ideas and trade-off descriptions come from work done by Snyder (1984).

A. DEFINITION

A flexible manufacturing system consists of a group of programmable production machines integrated with automated material handling equipment and under the direction of a central controller to produce a variety of parts at non-uniform production rates, batch sizes, and quantities.

B. SYSTEM CONCEPTS

Based on the flow of the workpieces through the FMS, three basic system concepts can be defined: sequential flow, single-station, and random flow. Each concept offers advantages and disadvantages and each offers differing degrees of flexibility and productivity.

C. SYSTEM COMPONENTS

The typical components that become the building blocks of the FMS used for material removal can be summarized as follows:

- Machine tools—machining centers, multiple-spindle machines, turning centers, and special-purpose machines.
- Material handling systems—industrial robots, tow-line carts, AGVs, conveyors, shuttle systems
- Computer(s)/controller(s)—minis, micros, PLCs
- Coordinate measuring machines (off-line quality assurance)
- Automatic gaging/compensation stations (on-line quality control)
- Automatic part wash stations

D. SYSTEM SOFTWARE

Computer software is used to assist in the design of the FMS, to control the use of the FMS resources, and to monitor the operation and performance of the FMS. As such, some software is designed to operate off-line while others provide real-time interaction with vital FMS functions. The FMS software is available from many sources and can represent a substantial percentage of the investment in the FMS.

Typical functions performed by FMS software include

- Capacity planning
- Simulation
- Production scheduling
- Process planning
- Tool management
- Maintenance planning and reporting
- Production monitor and control
- Machine/process control
- Machine diagnostics

In the sections that follow, each software type will be examined.

II. System Issues

A. FLEXIBILITY TRADE-OFFS

Specification of an FMS involves careful consideration of product design and production operation issues. A number of factors influence both the type and degree of flexibility that are possible, and the factors will be evaluated differently for different business situations:

- Production volumes expected
- Anticipated design change
- Types and variety of parts to be produced
- Part tolerance and material requirements
- Required manufacturing operations
- Expected operating/schedule change
- Capital available for investment

Often, trade-offs among the factors must be considered. The type(s) of product and product life, and expected design change influence the FMS design issues. The expected changes in production schedule to accommodate customer demand

and manufacturing resource availability influence the operational issues. For a given capital investment, a trade-off among these issues exists. This means that an FMS that is specified to respond to needed changes in design may not respond well to needed changes in production schedule or model mix (see Figure 1).

1. Product Design Flexibility

Product design flexibility makes it possible to accommodate a variety of product designs, including new products and modified designs. Much of this capability stems from recent advances in manufacturing controls that allow design details to reside in a machine's software rather than the mechanics of its physical structure.

2. Operating Flexibility

Operating flexibility makes it possible to reduce manufacturing costs while responding to changes in manufacturing conditions, such as production schedules and model mix. An important feature of systems with a high degree of operating flexibility is their capability to make up lost production. In the past, operating flexibility was often achieved through large work-in-progress inventories—an approach that is no longer affordable.

3. Other Trade-Offs

Similar trade-offs exist between the capacity of a system and its capability to manufacture a variety of parts; between required investment for a system and the needed production volumes; and between dedication of the FMS to produce parts for a particular product and dedication of the FMS to produce a family of parts for several product models. The type and degree of flexibility that are

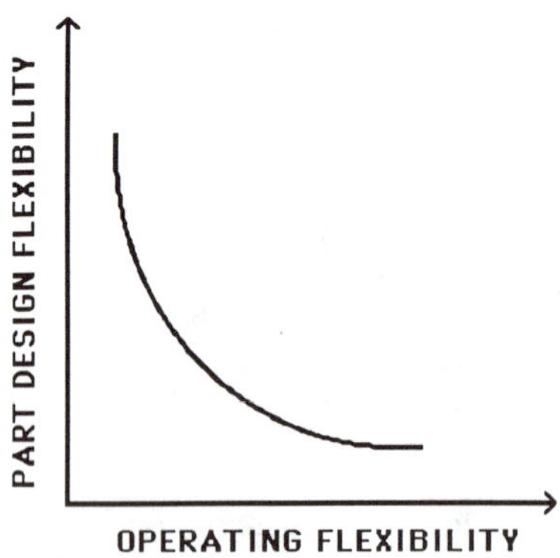

Figure 1. Trade-off between part design flexibility and operating flexibility.

derived from a careful analysis of the various trade-offs will directly affect the specification of the FMS concept and the level of capital investment that will be required.

The software products used to evaluate the various flexibility trade-offs are highly complex and in most cases proprietary to the companies supplying the FMS. Simulation models are often built to evaluate a specific FMS design concept and a model often must be rewritten if substantial changes to the system concept are to be considered.

B. SYSTEM CONCEPTS

Flexible manufacturing systems can be classified into three basic system types according to the flow of the workpieces through the system: sequential flow, single-station, and random flow. Each system concept represents a different capability for responding to changes in product design and production schedules.

1. Sequential Flow

A sequential flow system is one in which all workpieces follow the same path through the system. The sequential flow system might be characterized as a numerical control (NC) transfer line. Obviously, the sequential flow system favors the situation where all workpieces belong to a well-defined family of parts with each part in the family requiring the same basic operation sequence. If certain operations on a part must be repeated in the sequence of operations, it may be necessary to duplicate a particular machine at another location in the line. Since not all parts in the part family selected for the system require the full capability of each machine in the line, a high probability of over- and underutilized machines exists. The sequential flow system tends to respond better to operating change than to product design change.

2. Single-Station

A single-station system is one in which the workpieces do not move through the system. Instead, a large variety of tools (either single spindle or multiple spindle) are delivered to the workpiece at the single-station machine. The concept can be designed to favor design flexibility or operating flexibility. Machining heads can be added or modified to change the operations. While seemingly limited in production capacity, the single-station system can achieve higher levels of productivity by increasing the number of machining stations and using the same tooling among the stations. The tooling can be delivered to the machining station in a sequential or random order and redirected for maintenance or change-over, as necessary. The type of flexibility provided will depend on the scheduling scheme used.

3. Random Flow

A random flow system is one in which the material handling equipment can simultaneously move workpieces from any machine to any other machine in any sequence, thus producing several parts at the same time. The flow sequence depends on the specific operations required for the workpiece and the current availability of a machine that can perform the next operation. The material handling system of the random flow concept tends to be complex, but it can be justified when workpieces of entirely different configuration and different operation sequences must be processed. To improve balancing of the part flow, it may be necessary to take out underutilized machines or add machines to relieve bottleneck operations. The random flow concept tends to respond better to product design change than to operating change.

III. Types of Software

Software for FMS can be divided into three broad categories: design, extrinsic functions, and intrinsic functions. The design software is used to identify the major components of the system concept, to evaluate design sensitivities and trade-offs for a given concept, and to demonstrate that the system is capable of meeting the planned production requirements.

Some of the extrinsic and intrinsic functions that support the FMS are displayed in Figure 2. Software for the extrinsic functions is used to plan and control the functions that take place outside the physical boundaries of the FMS. Software for the intrinsic functions is used to load and control the components within the physical boundaries of the FMS.

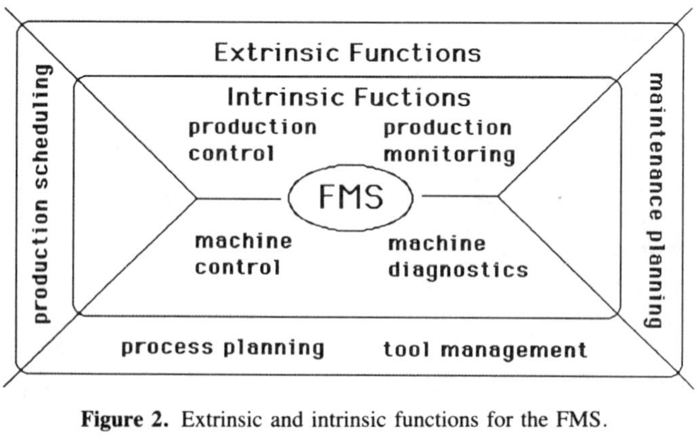

Figure 2. Extrinsic and intrinsic functions for the FMS.

A. DESIGN FUNCTIONS

1. Capacity Planning

Capacity planning is concerned with planning the needed manpower and machine resources to meet the master schedule during a specified time frame. The capacity of a system determines the amount of product that can be produced in a given time period. Capacity planning also is considered to be a refinement procedure wherein the long-range schedule is adjusted to the manufacturing needs of the nearer future (Rembold *et al.*, 1985). The planning horizon generally is one or several months and the time increments are weeks. The output of capacity planning can be an input to material requirements planning and other scheduling activities.

Role of Software

Capacity planning software is used to evaluate the impact of specified production levels on the resources provided through the FMS. It can be used to establish limits for the quantities of the various parts planned for the FMS and the limits on the mix of these quantities for the available machines, tooling and fixturing, and part pallets. It also can be used to evaluate the impact of introducing new part numbers into the FMS and identifying the changes that may be needed to accommodate the additional load.

Considerations

Capacity planning software can be used to evaluate factors that influence the capacity requirements for the FMS—factors such as the following:

- Minimum system requirements for part family demand
- Effect of change in part mix and demand levels
- Effect of change in planning horizon
- Effect on schedule of down machines or tool shortage
- Labor support required for various schedules (e.g., part programming, load/unload, tool replacement, setup, inspection, and maintenance)
- Minimum number of shifts required to meet schedule
- Capability to make up lost production

2. Simulation

Simulation has become an important tool when new manufacturing processes are conceived or existing ones are to be altered. A simulation can be described as a software tool that consists of methods, activities, and resources (Rembold *et al.*, 1985). A simulation is a model of a manufacturing process or system that can be used to emulate it. The model characterizes the layout configuration, the number of pallet shuttles in the system, and the production scheduling rules (Groover, 1987).

A simulation model for an FMS can be conceived as a four-level hierarchy: basic components, machine control, scheduling, and order processing and planning (Rembold *et al.*, 1985). The "basic components" are found at the lowest level of the hierarchy and consist of the machine tools, material handling equipment, inspection systems, and support equipment. The primary action at the basic components level is to implement instructions received from the next level.

At the machine control level, data are gathered and instructions are implemented on the basis of higher level goals and current events. Communication between machines and equipment at this level is limited.

At the scheduling level, future actions are planned and supervisory decisions are made. Primary and secondary process routings are determined for each part. When problems occur, rescheduling is done and new instructions are issued to continue operation of the FMS.

At the order processing and planning level, orders are generated and processed. Order status is maintained and operational statistics are calculated (e.g., processing times, equipment utilization, buffer use, tool changes).

Role of Software

The five main purposes of using simulation for FMS are to evaluate capacity and equipment utilization of the system, to identify the bottlenecks in the system, to compare the performance of alternative designs, to ensure that no fundamental weaknesses exist in the FMS design, and to develop operating strategies for work scheduling and job sequencing (Carrie, 1988). The simulation experiment may help answer questions such as those concerning system reserves, sensitivity to failure, wear, and process alternatives. Experiments also may be conducted to search for a more efficient layout, to test the effect of component failures, and to observe traffic congestion with respect to the number of part carriers and buffer sizes (Rembold *et al.*, 1985).

Some of the available simulation languages used to model FMS include GPSS/H (Wolverine Software Corp.), SIMAN (System Modeling Corp.), and SLAM II and MAP/1 (Pritsker & Associates, Inc.). Examples of FMS simulation models are documented in Rolston (1984) and Schriber (1984). A comparison of types and capabilities of eight FMS simulation systems is given by Bevans (1982).

Recent work is aimed at combining expert system and simulation methodologies to provide a means of dealing with the broad continuum of decisions that must be made to achieve a high level of productivity in an FMS. A conceptual model for an intelligent simulator design is given by Guariso *et al.* (1989) and by Bu-Hulaiga and Chakravarty (1988).

Considerations

Simulation is not an exact method. It is affected by the level of detail used, assumptions made, constraints specified, and probable error sources. The capability and quality of the simulation are determined by the following considerations:

- Data collection (including knowledge from people who know the system and potential problems)
- Simulation system architecture, performance measures, priority rules allowed, and utilization trade-off capability
- Representation of machine and material handling equipment characteristics, including carrier travel speed
- Representation of traffic rules (e.g., load parts on pallet fixture, launch loaded pallets into system, and resolve traffic conflicts)
- Quantitative skills of individuals writing/using simulation
- Reasonable design of simulation experiment(s)
- Reasonable presentation of data (considering performance measurement, correct interpretation of accuracy, and effective use of graphics)
- Documentation of the simulation (including user learning)

3. Knowledge-Based Systems

Both expert systems and knowledge-based systems are being used to analyze flexible manufacturing systems. An illustration of the architecture for a knowledge-based simulator is given by the Society of Manufacturing Engineers (SME) (1989). Kusiak (1987) reviewed relevant literature on expert and knowledge-based system developments that support FMS. The expert system incorporates a broad analysis and justification of the decision, while knowledge-based systems employ human knowledge to solve problems that ordinarily require human intelligence (Barr and Feigenbaum, 1981, see Kusiak, 1987). Expert systems (ES) have been applied to both design and operation of FMS. Expert systems can use data and constraints from the problem environment and generate a solution (stand-alone ES) or work jointly with a model and an algorithm (tandem ES). In addition to the literature review, Kusiak also provides examples for the machine layout problem and the scheduling problem.

A blackboard model-based expert system also can be used in the design of an FMS (Vujošević, 1990). The concept is based on work done on speech understanding by Fennel and Lesser (1977) and Erman *et al.* (1980). The main idea is to provide a kind of distributed intelligence for the various FMS design activities. The blackboard data structure serves as a communication interface among the various knowledge sources. To complete the expert system, a user interface and a control mechanism are needed.

Role of Software

Expert systems and knowledge-based systems provide a way to include knowledge sources in the decisions that are made in the FMS design process. Examples of these knowledge sources include characterizing part families, selecting equipment, layout of machines, building FMS models, analyzing output results from FMS simulation, and justifying capital investment.

Considerations

- Effective user interface (e.g., data entry, initial FMS layout, access existing models and data, constraint analysis)
- Definition and modification of part family
- Appropriate representation of system components and constraints
- Establish machine layout alternatives
- Relationship to other software planning systems (e.g., model definition, simulator input/output, data analysis/interpretation)
- Nature of command structure or control methodology
- Trade-off between degree of accuracy and software complexity

B. EXTRINSIC OPERATING FUNCTIONS

1. Production Scheduling

The master production schedule gives the timetable for end-product deliveries. The schedule is translated into material and component requirements by using material requirements planning (MRP) in conjunction with factory capacity planning. An operation schedule is developed to assign specific jobs to specific machines to meet the daily or even hourly requirements. The end products specified in the master schedule consist of components, each of which is manufactured by a sequence of processing operations. Operation scheduling involves the assignment of start dates and completion dates to the batches of individual components and the designation of machines on which the work is to be or at least can be performed. The scheduling problem is complicated by the fact that many parts may be competing for time on a fixed number of machines. The complications becomes more severe when interruptions and delays occur, for example, machine breakdowns, changes in job priority, and other operational problems (Groover and Zimmers, 1984).

Applications are also being developed using the expert system methodologies (Kanet and Adelsberger, 1985; Kusiak 1987). Other opportunities exist through integration of scheduling algorithms and simulation models in order to investigate trade-offs and do sensitivity analysis.

Role of Software

A variety of computer-based scheduling methods can be used in production. Selection of the best method will depend on the overall production objectives and the availability of resources. An economic analysis of the production run and economical batch sizes should be conducted. Given decisions about the number of parts needed and their due dates, the scheduling software determines the appropriate sequence to process a part on the available machine(s) in accordance with a predetermined set of priority rules. The allocation of parts to machines is referred to as machine loading and will depend on the operations to be performed, the capability of each available machine, and the priority scheduling rules.

Considerations

Selection of the operation scheduling software will depend on the types of parts produced, the type of FMS, and the scheduling priority system. Specific capabilities of the software to be considered include

- Priority rules allowed (e.g., customer order, job/operation precedence, work-center loading, job dispatch, and time conversion)
- Capability to deal with contingencies (e.g., machine breakdown, tool breakage, labor changes)
- Capability to optimize schedule variables to meet a production objective (e.g., minimum production cost)

2. Process Planning

Process planning represents the link between engineering design and shop floor manufacturing. It determines the manufacturing operations required to transform a part from a rough state to the finish state specified on the engineering drawing (Lindberg, 1990). More specifically, it is "that function within a manufacturing facility that establishes the processes and process parameters to be used in order to convert a piece-part from its original form to a final form that is predetermined on a detailed engineering drawing" (Niebel *et al.,* 1989). For machining, process planning includes determination of the specific sequence of machining operations and the selection of depth of cut, feed rate, and cutting speed for each cut of each operation. The machining parameters are in turn included within the CNC part program for the machine(s) that will cut the part. In some cases, the selection of tooling, fixturing, and inspection equipment also is included in the process planning for a part. Many versions of computer-aided process planning systems are available to assist the process planner, though none have been specifically included in software for FMS.

Role of Software

Process planning software is based on a "table lookup" (or database) structure, empirical calculations, or both. In some cases, specific data on machine tool capabilities and tooling also are included. Ultimately, the software generates a form of process plan or "route sheet." Integrated software that provides for the calculation of optimal machining parameters while the CNC part program is developed is not available. Either default values presented by the part programming software are accepted, or engineers enter parameters from tables or other calculation programs.

Considerations

Given the part design information, including production tolerance and surface finish specifications, the following activities can be performed by the process planner with the aid of software:

- Selection of workpiece material parameters based on design requirements
- Sequencing of operations (e.g., which operations and in what order)
- Grouping operations and part routing (considers which machines can do the needed single or multiple operations; also time standards)
- Selection of cutting tools
- Selection of machine tools
- Determination of machining parameters
- Specification of work standards
- Editing of process or tooling data
- Selection of holding devices (i.e., fixtures and jigs)
- Selection of inspection instruments

3. Tool Management

Because most of the value-added production tasks have been automated in an FMS, less need exists for a machine operator to be present at all times. Unless an effective tool management program is included in the overall management of the FMS, undue delay can result when the proper tool is not available or when tool failure stops production. Tool inventory levels must be maintained correctly to provide timely response to production needs, but avoid overstocking. A description of the elements to include in developing a tool management strategy is given by Gruver and Senninger (1988).

Role of Software

The tool management software must be able to identify and track tooling. Status of the tooling and inventory records must be maintained to ensure that the right tooling is available when it is needed. The system should allow the implementation of a tool replacement strategy consistent with the overall production objectives for the FMS. Interfacing should be provided to report data from the system and to receive data from the scheduling and process planning software. A sample listing of tool management software types and suppliers is provided in "American Machinist and Automated Manufacturing" (Directory of Software Suppliers, 1986).

Considerations

The requirements for a tool management software package are extensive (Rockwell/NASA Report 1980). Some of the key functional considerations for an effective software package follow.

- Assign tool identification number and maintain inventory records
- Maintain data on each tool (e.g., specifications and setup details)
- Accommodate machine tool relationship data
- Schedule changeover and maintenance of tooling
- Maintain and report status of tooling for each machine in the FMS
- Maintain records on tool failure
- Interface to peripherals and local area networks

4. Maintenance Planning/Reporting

Maintenance management plays a key role in maximizing the availability of machines for production. It is important to have highly reliable equipment for high productivity, but it also is important to prevent breakdown of equipment and repair equipment quickly when it does break down so production can continue.

Role of Software

A computer-based maintenance management system is used to schedule repair maintenance or preventive maintenance, issue maintenance reports, maintain materials inventory, release purchase orders for new maintenance materials, and support real-time supervision of machine components (Rembold *et al.*, 1985). Information on each piece of equipment is stored and made available for a wide variety of planning tasks and management reports. The software must be capable of generating work orders for inspection, replacement, or routine jobs on various frequency schedules. The system also must track the status of maintenance jobs and determine assignment of crew personnel.

Considerations

An example of the functional specifications for a maintenance management system is given in Bagadia (1988). Some highlights from the reference follow.

- Maintain equipment data (e.g., equipment number, name, location, cost, installation date, manufacturer information, vendor information, and spare parts information)
- Issue and track work orders (including craft availability, priority, and date needed)
- Maintain historical data (e.g., work-order number, date completed, material cost, labor cost, and reason for failure)
- Maintain stores inventory and appropriate charge-out for work orders
- Maintain vendor data and issue requisitions to replenish stores
- Interface to peripherals and local area networks

C. INTRINSIC OPERATING FUNCTIONS

1. Production Control

Production (or shop floor) control is concerned with the release of the production parts into the FMS, controlling the progress of the parts in process through the machine tools and inspection stations that are identified by the operation schedule, and acquiring current information on the status of each part in the system. When provided with the capability to determine optimal routing within the FMS, alternative assignments of jobs to machines can be determined based on the machine capabilities, part assignments, and utilization of the machines at any given time. When appropriate, the part program for the part assigned to a machine must be identified and downloaded to the machine. The status of part carriers in the system must be monitored and used to determine part load and unload requirements. Each part entering the FMS must be identified and production data associated with it must be accumulated as the part moves through the FMS.

Role of Software

The shop floor control software assigns parts to machines and tracks all information about the parts as they move through the FMS. On-line information is available by job or by work center.

Considerations

Selection of the production control software will depend on the types of parts produced, the capabilities of the machine tools, the tooling available at each machine, and the material handling system. Specific capabilities of the software include the following:

- Initiate system startup and shutdown procedures
- Establish (and reorder) material routing and report status

- Download CNC part programs when needed
- Maintain data on material handling system characteristics
- Adjust material flow to meet schedule and reduce congestion

2. Production Monitoring/Reporting

Production monitoring and reporting is concerned with data collection and management reporting. Records of the number of completed parts (both accepted and rejected), inspection results, tool change data, machine utilization, and other management data are collected and reported by the production monitoring/reporting system.

Role of Software

Production monitoring/reporting software provides standard and custom reports for managing the FMS resources. Problems are reported immediately to minimize delays and improve utilization.

Considerations

- Track use of labor, material (inventory), and machines versus production time
- Monitor completion of operations on each machine
- Perform special functions (e.g., statistical quality control calculations)
- Report on productivity measures, parts manufactured, quality performance, scrap/rework accounting, uptime/availability

3. Machine/Process Control

At the lowest level of the communication hierarchy for an FMS is the control of the manufacturing processes, operations, and material handling equipment. Parts are moved into and out of each scheduled machine or workstation. The dedicated material handling equipment picks up and delivers parts on command from the FMS computer. Each machine tool in the system is capable of starting, completing, and monitoring the machining operations for each part routed to it. Appropriate sensors and control algorithms are included to correctly position parts at the workstations and to provide process control.

Role of Software

Machine/process control software provides both control and monitoring capability. Part programs and data are stored at each machine tool. Tool wear and replacement strategies are integrated with the machine control software. Software may contain capabilities to adjust the process in real time for variations in process variables.

Considerations

Because of the complexity of the machine control software, it is important that the software be modular and flexible. Changes needed to control the equipment that result from design or schedule changes should be easily implemented. Additional considerations for the software include

- Graphics-based part programming capability (e.g., DNC, CNC, MDI)
- Identification of workpiece, pallet, fixture, and transport
- Control of material handling equipment (e.g., conveyors, AGVs, two-line carts, shuttle systems, industrial robots)
- Equipment synchronization
- Monitoring of tool wear and other sensors
- Support programs and data for operator

4. Machine Diagnostics

Because of the capital investment represented by an FMS, it is essential that the machines be available as much of the time as possible. Machine diagnostics provide the capability to locate malfunctions and provide maintenance personnel with the information they need to quickly restore the operation of a machine. A brief review of the sensor technology that supports machine diagnostics alternatives is given by Ball (1989).

Role of Software

Machine diagnostics software identifies a malfunction, the most probable reason for the malfunction, and solution alternatives for the malfunction. Some software includes the capability to alert maintenance personnel of an imminent failure, so corrective action can be initiated before the failure stops production. An advanced capability of machine diagnostics software switches control of failed FMS functions to a "hot" backup (redundant) control system or system component.

Considerations

The capability of the machine diagnostics software is related directly to the type and sophistication of sensors present in the FMS. Real-time monitoring is essential and "look-ahead" capability is desirable. Other considerations include the following:

- Maintain database on down occurrences
- Annunciate delay problems and potential delay problems
- Report functional errors
- Communicate with maintenance reporting system

D. RANDOM FLOW EXAMPLE

An excellent example of a large random flow FMS is installed at the John Deere Waterloo Works in Waterloo, Iowa. The system makes transmission cases and clutch housings for agriculture tractors and consists of the following equipment:

- Eleven machining centers, each with a 69-tool magazine
- Five multiple spindle head indexers
- Thirty-seven tow-line carts
- Central computer with hot backup
- Wash module
- Off-line coordinate measuring machine
- Electromechanical spindle probes for locating reference surfaces
- Tool life monitoring
- Centralized chip handling and coolant reconditioning system

The FMS makes eight different part numbers and accommodates design change through tooling and software changes. The system's second-generation software is written in Pascal and runs on a DEC 11/44 computer. Eight basic software tasks: router, scheduler, mover, traffic, management information, logger, operations, and file transfer help manage the system's resources to randomly manufacture the desired production part mix. Details about each of these functional software systems can be obtained from the manufacturer (Kearney & Trecker Corporation).

IV. Software Specification/Selection

A. SOFTWARE DESIGN REQUIREMENTS

For software to be effective, it must meet the requirements that are defined based on its purpose and how well it does its intended job. The requirements can be expressed in the form of absolute and functional specifications. The specifications provide a basis for detailing requirements for the software and can be used as a checklist against which various alternatives can be evaluated.

1. Absolute Specifications

In outline form, the absolute specifications include

a. Detailed specification, including

1. FMS characteristics
2. Program code

3. Computer hardware
4. Production capacity (production rate (volume) versus part mix)

b. Effective user interface
1. Minimal training requirements
2. Multiuser capability
3. Fast execution/response time

c. Software design
1. Modular structure
2. Design for growth (use existing data and upward compatible)
3. Interface to other software (other applications and management information software)

d. Field verification and test

e. Effective documentation (consistent with software version; tailored to user; easy to find information; clear, concise writing style; ample figures and tables; troubleshooting guide; flowcharts and variable dictionary)
1. Training manual/tutorial
2. User reference
3. Programmer reference

2. Functional Specifications

Considerations for the functional specifications include

a. Deliverables versus cost
1. Purchase package (basic package and optional packages)
2. Tech support and software service
3. Availability of source code
4. Upward compatible updates and enhancements
5. Documentation
6. Training
7. Hardware and support equipment

b. Quality
1. Vendor credibility (including on-time delivery)
2. Reliability

c. Programming
1. Program language (high level versus assembly; standard versus nonstandard)
2. Structure of program
3. Error recovery
4. Security
5. Diagnostic aids
6. Access to database

7. Interfaces to other applications
8. Graphics display

d. Control applications
1. CNC part program
2. Traffic control
3. Shuttle control
4. Tool control (including wear monitor and replacement transactions)
5. Quality control

e. Management applications
1. Scheduling
2. Production control
3. System monitoring

f. Reports
1. Management
2. Operational

B. DOCUMENTATION

An important consideration in the selection of a software product to support the FMS is the quality of the documentation that comes with the software. Four criteria can be used to judge the quality of documentation: availability, suitability, accessibility, and readability (Weiss, 1985).

1. Availability

Is documentation available? Some software products come with little or no formal documentation. While this practice may be acceptable for simple programs, the *need* for a high-quality user's manual should be satisfied by the supplier.

2. Suitability

What information does the user need? Obviously, the answer to this question depends on who the user is. A common mistake made by suppliers of software is to put the entire documentation into a single (often unwieldy or even unusable) manual. After a careful analysis of each user's needs, the developer of the documentation should "align each manual's content with the tasks and interests of particular readers."

3. Accessibility

Can the user find the needed information? Perhaps the documentation contains the information that answers the user's questions, but can the user find the information in an acceptable way? If the paths through the documentation are complex and cumbersome, even a skillful user may get lost. "User manuals that are both suitable and accessible (also tested and debugged) are likely to be called *task-oriented*. The task-oriented manual is one that contains only what the user needs, and in the most useful sequence."

4. Readability

How easily and accurately can the documentation be understood? User manuals of the highest quality require professional editing. The language used should be clear, concise, and communicate in a way that the user can gain the greatest understanding in the least amount of time. The reader should not have to read and reread (. . . and reread) the material to gain the needed understanding, often stumbling over language and editorial errors.

C. SOFTWARE SOURCE ALTERNATIVES

Software systems for FMS are obtained from three basic sources: a vendor, a consultant, or an in-house development. Each alternative has both advantages and disadvantages.

1. Vendor

A qualified vendor, often the hardware supplier, can supply software that has been field-tested and that can be upgraded as improvements and enhancements (both in software and in the FMS hardware) become available. The vendor also provides technical support to train personnel to use the software and to correct problems that arise during use of the software. Vendor-supplied software is expensive and usually cannot be modified by the user.

2. Consultant

Software consultants can provide highly customized software that can be changed easily as needs change. Since the software is usually a one-of-a-kind version and does not have the benefit of field testing, the software requires careful evaluation and delays may occur while the "glitches" get resolved. The risk can be high for consultant-developed software because effective software requires expertise in both software and application engineering. As a minimum, in-house application engineers should work closely with the consultant from the start of the project. While consultant-developed software also is expensive, costs can be controlled on a pay-as-you-go basis.

3. In-House Development

Software development by in-house professionals can take maximum advantage of the company's application engineering expertise. Since company

TABLE 1
Factors That Influence the Source Alternative Selected

Source	Flexibility (including upgrade)	Technical support	Risk	Lead time	Cost
Vendor	Limited	Very good	Low	Short	High
Consultant	Moderate	Good	High	Moderate	Moderate
In-house	High	Excellent	Moderate	Long	High

personnel develop the software, technical support is readily available to provide training, to resolve problems that arise, and to enhance the software when changes are needed. Software developed in-house can be designed upfront to interface with related application software and a management information system. Development times can be long (perhaps years), because the experts also are typically involved in other projects while doing the development work. On the other hand, implementation time can be short, because the people involved in the implementation are the same people involved in the development. Like the other alternatives, in-house developed software also can be expensive, because of the higher salaries and benefits that are paid to the team of experts required to do the development work.

4. Selection Factors

The software source alternative that is selected will depend on many factors and assumptions. A brief summary of key factors to consider and comparisons for the alternatives are given in Table 1.

D. SOFTWARE JUSTIFICATION

Obviously, the first requisite for computer software is need. However, using computer software to improve efficiencies and productivity will not happen if the operations supported are not first simplified and properly organized. *Automating inefficiency simply yields automated inefficiency*!

Like other investments or expenses, the expenditure for computer software requires financial justification. Reasons for acquiring the software can be identified that will produce (tangible) hard benefits, but other reasons produce only (intangible) soft benefits. Because the soft benefits that come from the use of computer software rarely can be quantified, traditional cost-justification procedures generally are inadequate for FMS software. Without considering the intangible benefits, the calculated return on investment from the purchase of software cannot compete with other investment opportunities.

TABLE 2
Reasons for Acquiring Tool Management Software

Business function	Reason
Finance	High tool purchase cost
Production control	Missed schedules
Engineering	Inefficient process planning
Quality	Nonconforming parts
Shop floor	Setup delays and part variability
Maintenance	Down machines

In most cases, the software is acquired for one of three reasons: to solve problems, to exercise control over some class of manufacturing variables, or to provide managers with information needed to make better decisions. Therefore, a good starting point is to delineate the reasons for acquiring the software as carefully as possible. The reasons then can be classified as hard or soft and appropriately used in the cost-justification process.

An effective procedure to identify the reasons for acquiring the software is "brainstorming." The group doing the brainstorming should represent the functional activities involved, for example, finance, production, engineering, quality, shop floor, maintenance, and other staff or shop personnel as appropriate to the need. The group, led by an effective facilitator, lists the specific problems that need to be solved, identifies the specific variables that need to be better controlled, and/or describes the specific information managers need for decision making.

An example of the product from a brainstorming session on software to support tool management is given in Table 2. Table 3 lists the hard and soft benefits that might be expected from use of the software.

Once the hard and soft benefits have been identified, quantitative and

TABLE 3
Hard and Soft Benefits from Tool Management Software

Hard benefits	Soft benefits
Reduce tool purchase cost	More reliable scheduling
Reduce scrap and rework cost	Improved process planning
	Reduced risk of defective parts
	Better process control
	Reduced risk of machine damage

qualitative values can be assigned to each benefit. A useful way to evaluate the soft benefits is a computer-based decision support system. The software product called Expert Choice (Crawford, 1984) is an example of such a system that can evaluate both quantitative and qualitative information about selection criteria and alternatives. Additional insights to consider when developing the economic justification for an FMS are given by Tipnis and Misal (1988).

E. SOFTWARE SUPPLIERS

Many software suppliers provide computer software packages for FMS or those that perform the tasks that support the operation of the FMS. The list of companies cited in Table 4 is not intended to be comprehensive but rather, a representative sample of the types of software available. Where appropriate, the products available from the companies are referenced to the three general types of software for FMS. Addresses for the companies are given in the Appendix.

To obtain the list given in Table 4, specific information about FMS software products was requested from 42 companies. Of those responding, only the 13 listed in Table 4 appeared appropriate for the FMS tasks. A listing of these and other software (and hardware) suppliers is given in the 1990 Automation Buyers Guide (1989).

TABLE 4
Software Suppliers

Supplier name	Design[a]	Extrinsic[b]	Intrinsic[c]
Ann Arbor Computer	II		2,3
ASK Computer Systems Inc.	I	A	1,2
Business Systems Design Inc.			2
COMPUMAX	I	A	2
Consilium	I,II	A	1,2
Ingersoll Milling Machine Company		A,D	1,2,3,4
Kearney & Trecker Corporation	I,II	A,B,C	1,2,3,4
Oliver Wight Companies		A	
PMX, Inc.			3
Production Systems Inc.	I	A	1,2
PROMIS Systems Corporation	I	A,C	1,2
Soft Systems Engineering, Inc.			2
Solid State Software, Inc.	I	A	1,2

[a]**Design:** (I) capacity planning, (II) simulation.

[b]**Extrinsic:** (A) production scheduling, (B) tool management, (C) maintenance planning/reporting.

[c]**Intrinsic:** (1) production (shop floor) control, (2) production monitoring/reporting, (3) machine/process control (including DNC/CNC), (4) machine diagnostics.

V. Conclusions/Trends

Flexibility in manufacturing systems provides an opportunity to capitalize on basic strengths of a company to ensure long-term leadership in manufacturing cost, products, and service to customers. It is important to keep in mind that flexibility comes in many types and degrees, and success will come only to the extent that the proper flexibility is specified for each situation. Implementation of flexible manufacturing systems involves development of various manufacturing concepts from which a few specific alternatives are selected, and then evaluated so that a single alternative can be selected and implemented.

As manufacturing moves increasingly toward systems such as FMS, it is critical that the requirements and opportunities of all phases of the business, including manufacturing, marketing, product design, and finance, be included in the development of concepts, selection of alternatives, and evaluation of the alternatives. The cost of capital equipment and its supporting software will be a major driving factor as well as a limiting factor in the ability of a company to remain competitive in a world market.

In addition to a very detailed business analysis, a comprehensive technical analysis of the system must be made. For example, the expanded manufacturing systems will incorporate new techniques and processes. In machining, spindle probes and spindle load monitoring are now common. Adaptive controls, tool life monitoring, broken tool detection, improved pallet positioning techniques, and spare tool selection all will come into common use in these large systems. Each technology brings with it the need for software to monitor and control the processes and operations.

The cost analysis of the various alternatives available will become more critical than in the past. In addition to the conventional cost areas, such as direct and indirect labor, inventory cost, scrap and rework cost, it will become increasingly necessary to consider cost impacts on other phases of the business. For example, consider the cost advantage of removing two years from a product introduction cycle that may be five to seven years into the future.

The sensitivity of the various alternatives to changing business conditions must be carefully evaluated. What if the forecasts for a particular product materialize faster or slower than anticipated? Or, how well do each of the FMS alternatives respond to a change in part design or production schedule?

Finally, since manufacturing systems will be provided by vendors with varying degrees of experience and capability, it is important to find and develop intelligent ways of working with vendors to develop basic concepts and turn these concepts into viable alternatives. Ways must be found to communicate the basic needs, goals, and strategies of the business to equipment and/or software suppliers in such a way that concepts directed toward the particular needs of the company can be developed.

The technology of manufacturing systems is expanding at an alarming rate. New concepts for processing contemporary materials are constantly becoming available. These concepts will require much more detailed analysis and their selection will require a careful blending of both technical and business skills.

Not to be overlooked is the fact that manufacturing systems represent a new challenge to manufacturing's ability to keep the systems running. The maintenance of these systems must be carefully specified and performed. Techniques of system management for the FMS resources must be developed. The relationship of the basic elements from a production control standpoint must be carefully investigated and designed. The FMS software products can provide the timely information needed for important management decisions and the control algorithms needed to get the greatest return on the shop floor.

Appendix: Software Supplier Addresses

Ann Arbor Computer Division of J. B. Webb Company, 1201 East Ellsworth, Ann Arbor, MI 48104

ASK Computer Systems, Inc. 2440 W. El Camino Real, P.O. Box 7640, Mountain View, CA 94022-7640

Business Systems Design, Inc. 1205 Wall Street, Oconomowoc, WI 53066

COMPUMAX P.O. Box 7239, Menlo Park, CA 94026

Consilium 640 Clyde Court, Mountain View, CA 94043

Ingersoll Milling Machine Company 707 Fulton Avenue, Rockford, IL 61101

Kearney & Trecker Corporation 11000 W. Theodore Trecker Way, Milwaukee, WI 53214-1127

Oliver Wight Companies 5 Oliver Wight Drive, Essex Junction, VT 05452-3499

PMX, Inc. 33129 Schoolcraft Road, Livonia, MI 48150

Production Systems, Inc. 526 South E Street, Santa Rosa, CA 95404

PROMIS Systems Corporation 4699 Old Ironsides Drive, Suite 300, Santa Clara, CA 95054-1824

Soft Systems Engineering, Inc. 413 Norway Street, York, PA 17403

Solid State Software, Inc. 1401 N. Jesse James Road, Excelsior Springs, MO 64024

References

Bagadia, K. (1988). "How to Select and Justify Manufacturing Software." Genesis Press, Brookfield, Wisconsin.

Ball, K. E. (1989). Diagnostics: The frontal assault on downtime. *Programmable Controls* April, p. 32–34.

Bevans, J. P. (1982). First, choose an FMS simulator. *Am. Mach.* **126**(5), 143–145.

Bu-Halaiga, M. I. and Chakravarty, A. K. (1988). An object-oriented knowledge representation for hierarchical real-time control of flexible manufacturing. *Int. J. Prod. Res.* **26**(5), 777–793.

Carrie, A. (1988). "Simulation of Manufacturing Systems." Wiley, New York.

Crawford, D. (1984). Software support for the intuitive thinker" (reprint from *Wall Street Computer Review) 1984 Dealers' Dig., Inc.* November.

Directory of Software Suppliers (1986). *Am. Mach. & Autom. Manuf.* May.

Erman, D. L., Hayes-Roth, I., Lesser, V., and Reddy, R. (1980). The HEARSAY-II speech understanding systems: Integrating knowledge to resolve uncertainty. *ACM Comput. Surv.* **12,** 213–253.

Fennel, R. D., and Lesser, V. (1977). Parallelism in AI problem solving: A case study of HEARSAY-II. *IEEE Trans. Comput.* **C-26**(2), 98–111.

Groover, M. P. (1987). "Automation, Production Systems, and Computer-Integrated Manufacturing." Prentice-Hall, Englewood Cliffs, New Jersey.

Groover, M. P., and Zimmers, E. W. (1984). "CAD/CAM: Computer-Aided Design and Manufacturing." Prentice-Hall, Englewood Cliffs, New Jersey.

Gruver, W. A., and Senninger, M. T. (1988). Tooling management in flexible manufacturing systems. *AUTOFACT '88 Conf. Proc.* Chicago, Illinois, October/November, p. 17-11–17-19.

Guariso, G., Hitz, M., and Werthner, H. (1989). An intelligent simulation model generator. *Simulation* August, p. 57–66.

Kanet, J. H., and Adelsberger, H. H. (1985). "Expert Systems in Production Scheduling." Department of Business Analysis and Research, Texas A&M University, College Station.

Kearney & Trecker Corporation. "Level II Control for K-T Flexible Automation Systems with Rail Guided Vehicle MHS," Company Report. Kearney & Trecker Corporation, Milwaukee, Wisconsin.

Kearney & Trecker Corporation. "Manufacturing Automation Projects," Company Report. Kearney & Trecker Corporation, Milwaukee, Wisconsin.

Knight, R. (1989). Manufacturing software is key to success. *Ind. Comput.* September/October, **8**(7), 63–68.

Kusiak, A. (1987). Artificial intelligence and operations research in flexible manufacturing systems. *INFOR J.* **25**(1), 7–10.

Kusiak, A., and Heragu, S. S. (1988). Computer integrated manufacturing: A structural perspective. *IEEE Network* **2**(3).

Lindberg, R. A. (1990). "Processes and Materials of Manufacture," 4th ed. Allyn & Bacon, Needham Heights, Massachusetts.

Mason, F. (1986). Computerized cutting–tool management. *Am. Mach. Autom. Manuf.* **130**(5), 105–120.

Mellichamp, M. J., and Wahab, F. A. A., (1987). An expert system for FMS design. *Simulation* **48**(5), 201–208.

Niebel, B. W., Draper, A. B., and Wysk, R. A. (1989). Modern Manufacturing Process Engineering. McGraw-Hill, New York.

1990 Automation Buyers Guide (1989). *Manuf. Eng.* December, p. 55–103.

Parker, G. M. (1986). Artificial intelligence and distributed processing: Key technologies for the next generation FMS. *CASA/SME FMS Conf.* March, Technical Paper MS86–168, p. 1–14.

Rembold, U., Blume, C., and Dillmann, R. (1985). "Computer-Integrated Manufacturing Technology and Systems." Dekker, New York.

Rockwell/NASA Report (1980). "Tool Crib Automation Identification & Data Collection System: Feasibility Study" (Final Report), NASA Special Task Assignment No, 138-R2, NASA/Rockwell

Contract NAS 8-27980, NASA Sub-task 4.0, Rocketdyne Division, Rockwell Industries Corporation.

Rolston, L. J. (1984). Modeling flexible manufacturing systems with MAP/1. *Proc. ORSA/TIMS Spec. Interest Conf. Flexible Manuf. Syst. 1st, 1984,* p. 199–204.

Schriber, T. J. (1984). The use of GPSS/H in modeling a typical flexible manufacturing system. *Proc. ORSA/TIMS Spec. Interest Conf. Flexible Manuf. Syst. 1st, 1984,* p. 168–182.

Schriber, T. J. (1987). The nature and role of simulation in the design of manufacturing systems, simulation in CIM—AI techniques. *Proc. Eur. Simul. Multiconf., 1987.*

Snyder, C. B. (1984). Personal Papers and Discussion, Deere & Company, Moline, Illinois.

Society of Manufacturing Engineers (SME) (1989). "Tools and Manufacturing Engineers Handbook—Desk Edition." SME, Dearborn, Michigan.

Sullivan, W. G., and LeClair, S. R. (1985). Justification of FMS using expert system technology. *CASA/SME Autofact 5 Conf. Proc., 1985.*

Tipnis, V. A., and Misal, A. C. (1988). Economics of flexible manufacturing systems. "Integrating the Automated Factory." Soc. Manuf. Eng., Dearborn, Michigan.

Vujošević, R. (1990). Object oriented visual interactive modeling and simulation of flexible manufacturing systems. MS Thesis, Vanderbilt University, Nashville, Tennessee.

Weiss, E. H. (1985). "How to Write a Usable User Manual." ISI Press, Philadelphia, Pennsylvania.

CHAPTER 4

Flexible Manufacturing Systems Simulation

Sadashiv Adiga and Maged Dessouky
Department of Industrial Engineering and Operations Research
University of California, Berkeley
Berkeley, California

I. Introduction

Computer simulation has become a very popular tool in the analysis and design of manufacturing systems (Law, 1986). In the simulation of an existing or proposed manufacturing system, the analyst builds a model of the system which captures the features to be studied. Some of the features of the system include jobs or products to be processed, material inputs, relevant processing variables, and resources utilized. The analyst then directs the model to mimic the behavior of the system under real or assumed conditions, and collects information to be used in computing the performance measures of the system.

Because a simulation model is basically a descriptive rather than a normative model, it can be made to incorporate features of a complex real system which cannot be easily built into an optimization model, such as its stochastic and dynamic behaviors. In addition, simulation is a design and analysis tool that allows the analyst to study the performance of a complex system over a long period of time in a compressed time span, without interference in the system operations. Because of these features, simulation has grown in popularity among academic researchers and industrial practitioners alike. Many manufacturing companies now require that simulation studies of proposed manufacturing facilities be performed prior to making a final decision on implementation (Grant, 1987).

Flexible manufacturing systems (FMSs) fill the gap between flow shops which manufacture a large quantity of similar products and job shops which produce a large diversity of products in small quantities. The FMSs are designed to produce moderate quantities, 200 to 20,000 units per year (Groover, 1980), of diverse products. They utilize general-purpose machines, mostly numerical controlled (NC). The NC machines are usually clustered in manufacturing cells and connected by automatic material handling equipment such as robots and automatic guided vehicles (AGVs).

In an FMS, a product undergoes a sequence of operations on one machine or a succession of machines. Changeovers on the machines from one product type to another are frequent. Recent trends in the design and use of FMSs aim at achieving very short changeover times and maintaining minimum work-in-process inventories. As such, they present highly challenging problems to FMS planners and analysts who have to choose among numerous alternatives of machine types, cell layouts, and job routings. Simulation is a very useful tool to be used as an aid in making such decisions.

The following example illustrates the complexity of the decisions involved in FMS operations and how simulation can assist in making these decisions. Job orders arrive at a manufacturing cell. Each job consists of a small lot of products that require the same order of processing operations (i.e., turning, drilling, milling, and grinding). Different jobs require a different order of processing operations. An operation may be performed on alternative machines which may be identical or different. Products move from one machine to another, with one or more processing operations performed on each machine visited.

In this environment, the following decisions need to be made: For each processing operation on a product, which machine should perform the operation, and for each machine, which product should it process next? These decisions are intertwined since the assignment of a product to a machine depends on the latter's availability. The selection of a product by a machine depends on the job's characteristics such as job priority, due date, and availability time. Also, the selection decisions need to take into account machine characteristics such as capacity and reliability. In addition, the material handling equipment to transport products from one machine to another and the job to move first must be selected.

Such a decision situation is extremely complex and is further complicated by the need to account for the work-in-process material between machines. Contingency plans must be in place to decide what happens when the buffer storage is at capacity. Should the material balk outside the system or block the progress of material in preceding stations?

All these decisions are made according to explicit or implicit rules. For example, the rules to determine which job to process next on a machine are referred to as *dispatching rules*. The dispatching rules include first-in-first-out (FIFO), earliest due date (EDD), shortest processing time (SPT), and more complex rules or algorithms. Research on scheduling jobs in an FMS environment is gathering momentum, but currently no algorithm exists that can automatically be applied to a given situation once its parameters are known. This leaves simulation as an appropriate alternative.

A simulation model of the system can be constructed and run under different dispatching rules. Records are kept on variables such as the time between arrival of jobs, the waiting time in queues, the time jobs spend in the system, and the time each machine is idle, blocked, or busy. These records are then used to compute system performance measures under each dispatching rule such as the mean number of products at each buffer (queue), the mean number of jobs in the system (work-in-process inventory), the mean daily production rate of each product type, the mean time each product spends in the system, machine utilization, and other relevant measures.

The purpose of this chapter is to introduce to the reader the various aspects of the use of simulation models to analyze FMSs. References to some of the past simulation studies as applied to FMSs are presented in Section II. Section III describes the steps involved in a simulation study with an example provided in Section IV. Section V reviews some of the available simulation software languages that could be used to model an FMS. The limitations of simulation are discussed in Section VI.

II. Applications of Simulation

In numerous studies, simulation has proved to be an effective technique for analyzing and evaluating an existing or proposed manufacturing system. Simulation studies during the early design phase of a manufacturing system implementation tend to focus on equipment and material handling requirements. The output of these simulation studies aids in the design of the system specifications.

Caruso and Dessouky (1987) developed a simulation model of a proposed FMS system that assembles instrument panels. The system consists of an automated storage and retrieval system feeding several robotic assembly cells. The results of the simulation model aid in the design of the system specifications. Godziela (1986) uses simulation to evaluate performance of a proposed system to support certain levels of production activity. Falkner and Garlid (1987) use simulation to justify implementation of an FMS.

Simulation models of existing systems frequently are used to test different control and dispatching strategies. For example, Chang *et al.* (1985) use simulation to evaluate different dispatching rules in an FMS. By carefully controlling the use of generated random numbers, the authors implement an appropriate experimental design to identify significant differences between the various dispatching rules. Vig *et al.* (1989) discuss an example containing both design and decision-making issues in an FMS.

Readers are referred to the *International Journal of Production Research* (published by Taylor & Francis Ltd.) for additional papers on this topic of using simulation to analyze FMS.

III. Simulation Process

The simulation modeling process involves several interwoven steps. These steps are

1. Stating the project's objective(s)
2. Building the abstract model
3. Software selection
4. Input analysis
5. Building the simulation model
6. Model verification
7. Model validation
8. Output analysis

A. STATING THE PROJECT'S OBJECTIVE(S)

Before the simulation model is built, a clear statement of objectives must be defined. The objective of the study determines the level of detail to include in the model. For example, if the objective is to determine the capacity of a manufacturing line, product aggregation may be appropriate. However, if the objective is to evaluate different production schedules, product aggregation is not appropriate.

If the study involves a comparison of different alternatives, it is necessary to specify the criteria for comparison and evaluation of the alternatives generated.

B. BUILDING THE ABSTRACT MODEL

A thorough understanding of system operations is required to abstract the real world into a model. Building an abstract model involves determining which operations are to be included in the model and the boundaries of the model. The level of detail included in the model is closely related to the objective(s) of the simulation study. Medeiros (1989) suggests developing a process flowchart to show the product flow.

There are several advantages of developing an abstract model before building the simulation model. As more complexity is added to the model, the simulation software that is currently being used may not be satisfactory. As a result, a new simulation model might have to be developed using a different simulation software tool. The abstract model can serve as the starting point for the new model. Conversely, if the simulation software tool has not yet been selected, the abstract model can aid in selecting the appropriate software package. Section V presents some commercially available simulation software tools.

C. INPUT ANALYSIS

Input analysis involves data collection and the selection of probability distribution(s). Reliable data in the form needed for simulation purposes are often not immediately available (Carson, 1986). Some methods of obtaining the required data are

1. Time studies
2. Examination of historical records
3. The client's or an expert's best guess
4. A vendor's claims

Since manufacturing systems contain many sources of variability (machine breakdowns, processing time, etc.), random samples from probability distributions need to be generated. If enough data are collected, statistical tests such as chi-square and goodness-of-fit can be performed to determine the appropriate input distribution. Superimposing a suggested input distribution on a histogram of the data aids in understanding the form of the input distribution. Figure 1 shows an exponential distribution with a mean of 1.43 h superimposed on a histogram of repair time data. As the diagram shows, the exponential distribution is a relatively good fit for the repair time data. The parameters of the distribution (mean, standard deviation, etc.) are estimated using the available data.

Some common distributions used to represent processing time are triangular and normal. For the triangular distribution, estimates of the minimum, most

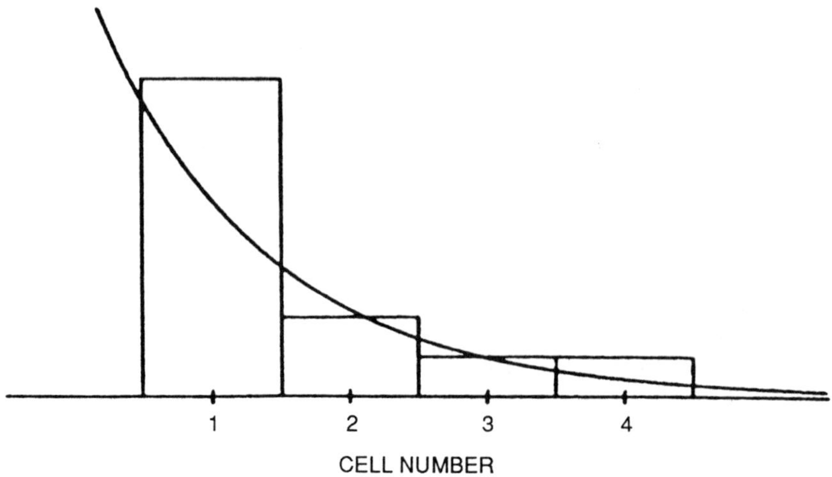

Figure 1. Histogram of repair time data.

likely, and maximum points need to be made. For the normal distribution, estimates of the mean and standard deviation are required.

D. BUILDING THE SIMULATION MODEL

The level of detail to include in the model is an important consideration in building a simulation model. The more detail included in the model, the more time that is required for data collection, programming, and computer processing time. Section IV illustrates building a simulation model using SLAM II (Pritsker, 1987) for an FMS.

E. MODEL VERIFICATION

Verifying the simulation model means to check if the simulation model is designed to the model specifications and is behaving as intended. Several techniques used to verify a simulation model are the trace output, graphical animation, and summary statistics.

The *trace output* is a listing of the events that are performed by the model. An *animation* is a graphical model of the simulation. It is generally easier to verify the model with a visual tool such as an animation than it is to look at a computer listing of the events (trace output). Developing an animation model adds an extra step to the simulation process. Since the animation is another model, the user needs to be certain that the errors visualized by the animation are due to the simulation model and not the animation model.

Another technique to verify the model is to look at the *summary statistics*. If they cannot be explained, there is something wrong with the model.

F. MODEL VALIDATION

Validating the model involves answering the question "Does the model accurately represent the real world?" Validity is obtained when the model output agrees with the input data (Sayers, 1989). If the system being modeled currently exists, the results of the simulation model and the real world can be compared. If the system does not exist, it is more difficult to validate the model. However, a domain expert can be asked to look at the simulation results to see if they make sense. The same tools used for model verification can be used for model validation.

One of the main problems facing a simulation modeler is gaining the client or end user's acceptance of the model's validity. Carson (1986) describes a credible model as one which is accepted by the client as being valid, and is used as an aid in making decisions. He suggests that involving the client or end user in a team effort makes it much more likely that the model will be accurate and used in the decision-making process.

G. OUTPUT ANALYSIS

Effectiveness of the simulation output analysis is governed by the design of the simulation experiment and the study's objective(s). Parameters that need to be determined include the number of runs, simulation run length, and initialization warm-up period.

Multiple simulation runs are made to estimate the variability of the output performance measures. The variability can be used to estimate statistical confidence intervals of the results. Readers are referred to standard textbooks in statistics or simulation for further details on calculating statistical confidence intervals. The number of runs to make depends on the constraints of the computer system and desired precision of the estimates. The constraints of the computer system include computer size, availability, and cost (Roberts, 1983). Roberts computes the percentage reduction in the half-width of a statistical confidence interval in terms of the standard deviation with each additional run. The results show that each additional run beyond eight or nine yields less than a 10% marginal reduction in the half-width.

The simulation run length is based on the time units of the processes that are being modeled. For example, if the time units are in minutes, then running the simulation for a month may be sufficient. Determining the run length is largely based on the modeler's judgment. If computer resource time is not an issue, it is best to run the simulation longer than necessary to ensure that the variance of the estimate is not large.

How to handle initial conditions depends on whether a transient or steady-state simulation analysis is being performed. A transient system is one that has clear beginning and ending points. For example, a bank opens and ends the day clear of customers. In a transient system, the simulation model is initialized to a given state.

A steady-state simulation analysis is concerned with the long-run behavior of the system. In most manufacturing lines the system is not cleared of parts at the end of the day. In these situations a steady-state analysis is appropriate.

In a steady-state analysis, the initial conditions of the model (warm-up period) do not matter since the analyst is concerned about estimates of the performance measures when the system reaches a steady state. The initial conditions should be removed from the analysis so as not to bias the results. This is accomplished by running the simulation for some length of time and clearing the statistics at that time.

The length of the warm-up period can be determined by plotting key output performance measures such as the number of parts waiting at the bottleneck machine(s) over time. A plot of the number of parts waiting to be picked by a stacker crane over time is shown in Figure 2. From the plot, the initialization warm-up period ends between 360 and 480 time units. To remove the initialization bias, the statistics should be cleared after 480 time units.

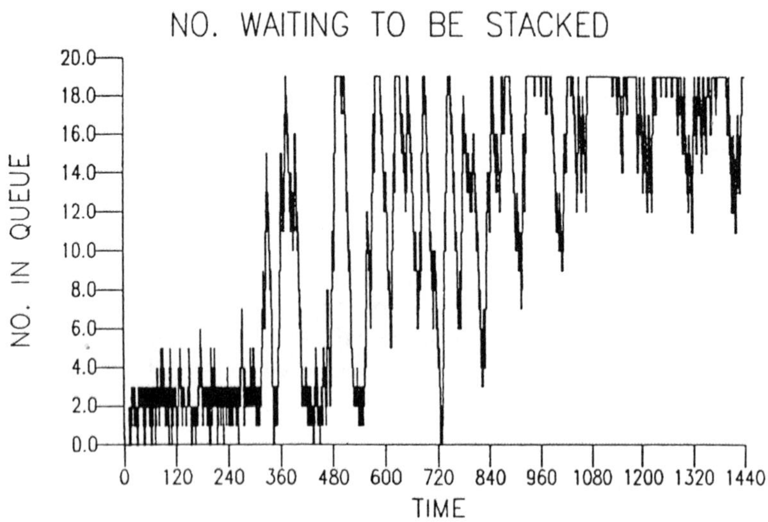

Figure 2. Number waiting to be stacked over time.

IV. A Model of a Flexible Manufacturing System

The example presented in this section is designed to illustrate some of the concepts involved in modeling modern FMSs.

A. STATEMENT OF OBJECTIVES

Company XYZ is considering investing in a manufacturing system. The decision whether to invest in the new system depends on the system's expected performance level. As a result, management has authorized a simulation study to predict the following statistics:

1. The mean number of jobs in the queue and the mean delay in the queue for each workstation
2. The mean time in the system for each job type
3. The proportion of time that machines are processing jobs, blocked, and idle for each machine group
4. The utilization of the transporter
5. The mean time a job waits for transporter pick-up

These statistics will be used to help management decide whether or not to invest in the system.

B. SYSTEM DESCRIPTION

A manufacturing shop consists of a receiving/shipping station and five workstations. workstations 1, 2, . . ., 5 consist of 3, 3, 4, 4, and 1 identical machines, respectively. The distances (in feet) between the six workstations (the receiving/shipping station is station 6) are shown in Table 1.

The time between arrival of jobs is an exponential random variable with a mean of 0.25 h. Jobs are of type 1, 2, or 3 with probabilities 0.3, 0.5, and

TABLE 1
The Distance Matrix

Workstation	1	2	3	4	5	6
1	0	90	100	180	200	270
2	90	0	100	200	180	270
3	100	100	0	100	100	180
4	180	200	100	0	90	100
5	200	180	100	90	0	100
6	270	270	180	100	100	0

TABLE 2
The Job Routings

Job type	Workstation routing
1	3, 1, 2, 5
2	4, 1, 3
3	2, 5, 1, 4, 3

0.2, respectively. Each job enters the system at the receiving station, travels to workstations on its route, and then leaves the system at the shipping station. Table 2 shows the job routings.

The processing time of a job at a workstation is a normal random variable whose mean depends on the job type and the workstation. The standard deviation of the processing time is 30% of the mean. The mean service time, in hours, of each job type at each workstation on its route is contained in Table 3. For example, job-type 2 has a mean service time of 1.10 h at workstation 4.

The dispatching rule at each workstation is FIFO. When a machine finishes processing a job, the job blocks the machine (the machine cannot process another job) until the job is removed by the transporter.

A job moves from one workstation to another by a transporter (e.g., an automatic guided vehicle). There is one transporter in the shop and it moves at a speed of 5 ft/sec. The transporter processes requests on a FIFO basis. When the transporter finishes moving a job to a workstation, it remains at that workstation if there are no pending job requests.

The shop runs one shift a day. Each shift is 6.5 manufacturing hours.

C. MODEL DEVELOPMENT

The simulation model is developed using SLAM II (see Appendix A for the model code). The entities of the system are the jobs. The workstations and transporter are modeled as resources.

To verify and validate the model, a trace of the simulation events is created

TABLE 3
Mean Service Time of Each Job at Each Workstation

Job type	Mean service time (h)
1	0.50, 0.60, 0.85, 0.50
2	1.10, 0.80, 0.75
3	1.20, 0.25, 0.70, 0.90, 1.00

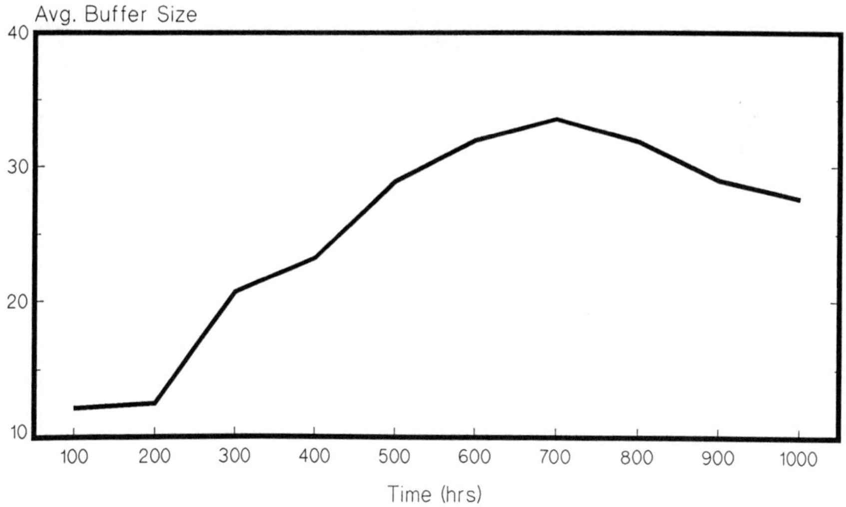

Figure 3. Average buffer size versus time (workstation 1).

(see Appendix B for a listing of the trace). Examination of the trace shows that the simulation model behaves according to the system description.

To determine the length of the initialization bias, the average number of jobs waiting at workstation 1 is plotted over time. Figure 3 shows this plot. From the diagram, the initialization bias is determined to be 500 h. Statistics are cleared after 500 h.

As suggested in Section III, eight simulation runs are made. Each simulation run is 200 days (1300 h) after statistics are cleared.

D. DISCUSSION OF RESULTS

The job statistics are contained in Table 4. Type-2 jobs have the highest average daily production rate of 13.10 jobs with a standard deviation of 0.21

TABLE 4
Job Statistics

Statistic	Job type 1	Job type 2	Job type 3
Average production rate (daily)	7.79	13.10	5.21
Standard deviation production rate (daily)	0.13	0.21	0.08
Average time in system (h)	8.20	7.60	10.00
Standard deviation time in system (h)	1.30	1.30	1.30

TABLE 5
Workstation Statistics

Statistic	Workstation number 1	2	3	4	5
Percentage busy	95.86	65.90	73.15	74.35	78.20
Percentage blocked	1.47	0.77	0.85	0.65	1.80
Average queue size	17.96	0.44	0.38	0.74	1.59
Standard deviation queue size	5.42	0.06	0.03	0.09	0.26
Average waiting time (h)	4.45	0.22	0.09	0.26	0.79
Standard deviation waiting time (h)	1.32	0.02	0.01	0.03	0.12

jobs. Type-3 jobs stay in the system the longest with an average time in the system of 10 h and a standard deviation of 1.3 h. A 95% confidence interval of time in the system from the eight simulation runs for type-3 jobs is (8.94, 11.06) h. The upper limit can be used to set delivery due dates for type-3 jobs.

The transporter is busy 29% of the time, and requests wait on the average 0.61 min for the transporter. These results indicate that one transporter is sufficient to meet the current demand.

Table 5 contains the workstation statistics. The results show that workstation 1 is the bottleneck station. It is busy processing jobs 95.86% of the time. All other workstations are busy around 70% of the time. The proportion of time any workstation is blocked is low because the transporter is underutilized.

Workstation 1 has on the average 17.96 jobs waiting to be processed with a standard deviation of 5.42 jobs. The jobs on the average wait 4.45 h at workstation 1. The other workstations have much lower queue sizes.

The overall average time in the system is 8.26 h. Waiting for workstation 1 constitutes about 54% of the time a job is in the system. This result suggests that another machine may be needed at workstation 1. If another machine is added, machine utilization at workstation 1 decreases to 72%. The average queue size and waiting time reduce to 0.563 jobs and 0.138 h, respectively. Thus, the overall average time in the system reduces to 3.93 h.

V. Simulation Software

Discrete event simulation can be implemented using most programming languages. But today most simulation models are written using special purpose languages that contain basic utilities needed to perform most simulations.

A. GENERAL-PURPOSE COMPUTER LANGUAGES

FORTRAN was the most popular language for simulation modeling in the 1970s because of its ability to handle numeric computations conveniently and efficiently. In recent years, PASCAL and C have emerged as competitors to FORTRAN. These relatively newer general-purpose languages provide data structures that make the abstraction process easier.

An advantage in using a general-purpose language is the ability to analyze a large number of problems since the user defines and develops all the model constructs. However, a major disadvantage is that a significant amount of computer code must be written to manage the simulation model and collect statistics. General-purpose simulation languages are designed to reduce the amount of user programming. Basic utilities are provided to manage the simulation model and collect statistics.

B. GENERAL-PURPOSE SIMULATION LANGUAGES

General-purpose simulation languages aid users by providing utilities that handle repetitive tasks such as initialization and time advancement. Some examples of general-purpose simulation languages are GPSS (Gordon, 1975), SLAM II (Pritsker, 1987), SIMSCRIPT (CACI, 1983), and SIMAN (Pegden, 1985). These simulation languages contain collections of FORTRAN subroutines with a set of control statement used to invoke them. The system-provided subroutines reduce the amount of code a modeler has to write when building a model.

Utilities offered by some of the popular general-purpose simulation languages include

1. Automatic queue management, including generation of statistics on average waiting time, maximum waiting time, and average queue length
2. A generator of uniformly distributed pseudorandom numbers used to support samples from probability distributions such as Normal and Poisson
3. Provisions to control the number of runs, the length of a simulation run, and time to clear statistics
4. Automatic generation of entity arrivals to the system
5. Collection of statistics (i.e., mean, standard deviation, minimum, maximum, etc.) for chosen variables
6. An optional animation capability

Most simulation languages have the basic constructs needed to represent common production facilities. Simple examples can be modeled easily in these languages but large applications involve complex development. As Grant (1987) points out, simulation models of production facilities are at times expensive to build and cumbersome to use on a daily basis. For this reason, research into the use of alternative approaches to simulation system design is being pursued. Knowledge-based systems (Shannon *et al.,* 1985; Ben-Arieh, 1986) and object-oriented programming systems (Adiga, 1989; Olumolade *et al.,* 1990) are examples of the new approaches.

C. SPECIAL-PURPOSE SIMULATION PACKAGES

Special-purpose simulation packages have been developed for modeling manufacturing systems. MAP/1 (Wortman and Miner, 1986) and SIMFACTORY (Klein, 1986) are two such packages. These packages use high-level building blocks to provide a user-friendly interface. XCELL (Jones and Maxwell, 1986) has a graphics interface. Building simulation models from these packages is easy and usually faster than using a general-purpose simulation language. However, these packages are designed to analyze only manufacturing systems and are limited to certain decision rules. Since the packages are fixed in the view of the world they take, using the package to model a different view may be difficult.

D. SELECTION OF SIMULATION SOFTWARE

Personal computers (PCs) have recently become attractive platforms for running FMS simulations due to the availability of large memory and 16-bit (or more) microprocessors. Most of the general-purpose simulation packages discussed earlier have PC versions of their mainframe software. These versions usually have additional interactive and graphics utilities.

A comprehensive catalog of simulation software appears in the October issue of *Simulation,* published by the Society of Computer Simulation. *Industrial Engineering* magazine, a publication of the Institute of Industrial Engineers, also publishes special issues devoted to simulation.

The selection of a language should be guided by the actual needs of users and the application under consideration. Some of the desirable features in simulation software are discussed by Haider and Banks (1986).

VI. Limitations of Simulation

Simulation modeling is not without its limitations. Since any model is an abstraction of reality, the modeler may ignore some of the details and focus only on those most relevant to the problem at hand. The results should be evaluated in the context of the assumptions made (or the details ignored).

Not all questions can be addressed by simulation. While it is a good tool for evaluating a number of alternative system designs, the modeler must generate the data for the alternatives (or come up with alternatives). Unlike optimization techniques, simulation does not pick an optimum solution to a problem.

The simulation study is a time-consuming process and requires the modeler to have skills in the following areas: data analysis, statistics, programming, and project management. Consequently, simulation is appropriate as an analysis tool when the system is too complex to be studied by performing a data analysis or by building an analytical model.

VII. Conclusion

During the design phase of an FMS, a simulation analysis can aid in evaluating the design. The results of the simulation model can aid in determining the required number and location of machines, the location and size of the buffers, and the appropriate type of material handling system.

After the FMS has been installed into the manufacturing environment, continuous improvement of the system is needed to fully utilize its resources and meet the production goals. For example, a simulation model can be used to analyze different scheduling strategies without disturbing system operation.

A wide range of simulation software tools are available to make the programming task easier. Selection of the appropriate software tool depends on the application. In building the simulation model, careful consideration must be given to model validity and statistical issues.

Appendix A: SLAM II Simulation Model Code

```
gen, dessouky, fms example, 10/31/89, 8, , n, , , , 72;
limits, 10, 8, 6000;
init, 0, 1800;
montr, clear, 500;
;
; time state statistics
;
timst, xx(1), workstation 1, 3/. 5/1;
timst, xx(2), workstation 2, 3/. 5/1;
timst, xx(3), workstation 3, 3/. 5/1;
timst, xx(4), workstation 4, 3/. 5/1;
timst, xx(5), workstation 5, 3/. 5/1;
;
; initialize the array
;
; setup distance matrix
;
array(1, 6)/0, 90, 100, 180, 200, 270;
array(2, 6)/90, 0, 100, 200, 180, 270;
array(3, 6)/100, 100, 0, 100, 100, 180;
```

Appendix A (continued)

```
array(4,6)/180,200,100,0,90,100;
array(5,6)/200,180,100,90,0,100;
array(6,6)/270,270,180,100,100,0;
;
; setup route matrix
;
array(7,5)/3,1,2,5,6;
array(8,4)/4,1,3,6;
array(9,6)/2,5,1,4,3,6;
;
; setup service time matrix
;
array(10,4)/.5,.6,.85,.5;
array(11,3)/1.1,.8,.75;
array(12,5)/1.2,.25,.70,.90,1.00;
;;;;;;;;;;;;;;;;;;;;;;;;
;
; intlc statements that initialize
; the user input global variables
;
;;;;;;;;;;;;;;;;;;;;;;;;;
intlc,xx(9)=6;
network;
;
;
; slam network model of fms
;
;
; resource statements
;
     resource/1,wkstat1(3),1;        workstation 1
     resource/2,wkstat2(3),2;        workstation 2
     resource/3,wkstat3(4),3;        workstation 3
     resource/4,wkstat4(4),4;        workstation 4
     resource/5,wkstat5(1),5;        workstation 5
     resource/6,recv(20),6;          receiving
     resource/7,trans(1),7;          transporter
;
;
;
; network flow
;
;
;
;
; job arrival
;
c1    create, expon(.25),,2,,1;                  create board arrival
w6    await(6),recv,,1;                          wait for input
      act,,.3,t1;                                type 1 board
      act,,.5,t2;                                type 2 board
      act,,.2,t3;                                type 3 board
```

```
;
t1      assign,atrib(1)=1,atrib(3)=6,atrib(5)=tnow;
                                                    go to wait for transporter
;
t2      assign,atrib(1)=2,atrib(3)=6,atrib(5)=tnow;
        act,,,tr;                                   go to wait for transporter
;
t3      assign,atrib(1)=3,atrib(3)=6,atrib(5)=tnow;
        act,,,tr;                                   go to wait for transporter
;
; transporter move logic
;
tr      await(7),trans;                             wait for transporter
        act/1,.0000556*array(xx(9),atrib(3));       go to job pickup
        colct,int(5),wait for trans;                collect stat on transporter wait
        assign,ii=atrib(3),xx(ii)=0;                set machine state to idle
        free,atrib(3);                              free machine (unblock)
        assign,atrib(4)=atrib(4)+1,
                xx(7)=atrib(3),
                ii=atrib(1)+6,
                atrib(3)=array(ii,atrib(4)),
                xx(9)=atrib(3);
        act/2,.0000556*array(xx(7),atrib(3));       delivery job to desired location
        free,trans,1;                               free transporter
        act,,atrib(3).eq.6.and.atrib(1).eq.1,trml; go to shipping
        act,,atrib(3).eq.6.and.atrib(1).eq.2,trm2; go to shipping
        act,,atrib(3).eq.6.and.atrib(1).eq.3,trm3; go to shipping
        act/4,,atrib(3).ne.6,mach;                  go to machine processing
;
; machine processing logic
;
mach    await(atrib(3)=1,5),atrib(3);               wait for machine
        assign,ii=atrib(3),xx(ii)=1,
                ii=atrib(1)+9,
                xx(7)=array(ii,atrib(4)),
                xx(8)=.3*xx(7);                     set machine state to busy
        act/5,rnorm(xx(7),xx(8));                   process job
        assign,ii=atrib(3),xx(ii)=2,
                atrib(5)=tnow,1;                    set machine state to blocked
        act/6,,,tr;                                 go to wait for transporter
;
; done processing
;
trml    colct,int(2),time sys 1;                    collect time in system stat for job 1
        term;
;
trm2    colct,int(2),time sys 2;                    collect time in system stat for job 2
        term;
;
trm3    colct,int(2),time sys 3;                    collect time in system stat for job 3
;
; end of network
;
        end;
fin;
```

Appendix B: Trace from Simulation Run

SLAM II TRACE BEGINNING AT TNOW = .0000E+00

		NODE ARRIVAL				ACTIVITY SUMMARY		
TNOW	JEVNT	LABEL	TYPE	CUR ATRIB BUFFER		IND	DURATION	END ND
.000E+00		C1	CREATE	.000E+00	.000E+00			
				.000E+00	.000E+00			
				.000E+00	.000E+00			
				.000E+00	.000E+00			
		W6	AWAIT	.000E+00	.000E+00			
				.000E+00	.000E+00			
				.000E+00	.000E+00			
				.000E+00	.000E+00			
						0	NO RELEASE	T1
						0	.000	T2
		T2	ASSIGN	.000E+00	.000E+00			
				.000E+00	.000E+00			
				.000E+00	.000E+00			
				.000E+00	.000E+00			
						0	.000	TR
		TR	AWAIT	.200E+01	.000E+00			
				.600E+01	.000E+00			
				.000E+00	.000E+00			
				.000E+00	.000E+00			
						1	.000	
			COLCT	.200E+01	.000E+00			
				.600E+01	.000E+00			
				.000E+00	.000E+00			
				.000E+00	.000E+00			
			ASSIGN	.200E+01	.000E+00			
				.600E+01	.000E+00			
				.000E+00	.000E+00			
				.000E+00	.000E+00			
			FREE	.200E+01	.000E+00			
				.600E+01	.000E+00			
				.000E+00	.000E+00			
				.000E+00	.000E+00			
			ASSIGN	.200E+01	.000E+00			
				.600E+01	.000E+00			
				.000E+00	.000E+00			
				.000E+00	.000E+00			
						2	.006	
.556E-02			FREE	.200E+01	.000E+00			
				.400E+01	.100E+01			
				.000E+00	.000E+00			
				.000E+00	.000E+00			
						0	NO RELEASE	TRM1
						0	NO RELEASE	TRM2
						0	NO RELEASE	TRM3
						4	.000	MACH
		MACH	AWAIT	.200E+01	.000E+00			
				.400E+01	.100E+01			
				.000E+00	.000E+00			
				.000E+00	.000E+00			
			ASSIGN	.200E+01	.000E+00			
				.400E+01	.100E+01			
				.000E+00	.000E+00			
				.000E+00	.000E+00			
						5	1.070	
.299E+00		C1	CREATE	.000E+00	.299E+00			
				.000E+00	.000E+00			
				.000E+00	.000E+00			
				.000E+00	.000E+00			

TNOW	JEVNT	NODE ARRIVAL LABEL	TYPE	CUR ATRIB BUFFER		ACTIVITY SUMMARY IND	DURATION	END ND
		W6	AWAIT	.000E+00	.299E+00			
				.000E+00	.000E+00			
				.000E+00	.000E+00			
				.000E+00	.000E+00			
						0	NO RELEASE	T1
						0	NO RELEASE	T2
						0	.000	T3
		T3	ASSIGN	.000E+00	.299E+00			
				.000E+00	.000E+00			
				.000E+00	.000E+00			
				.000E+00	.000E+00			
						0	.000	TR
		TR	AWAIT	.300E+01	.299E+00			
				.600E+01	.000E+00			
				.299E+00	.000E+00			
				.000E+00	.000E+00			
						1	.006	
.304E+00			COLCT	.300E+01	.299E+00			
				.600E+01	.000E+00			
				.299E+00	.000E+00			
				.000E+00	.000E+00			
			ASSIGN	.300E+01	.299E+00			
				.600E+01	.000E+00			
				.299E+00	.000E+00			
				.000E+00	.000E+00			
			FREE	.300E+01	.299E+00			
				.600E+01	.000E+00			
				.299E+00	.000E+00			
				.000E+00	.000E+00			
			ASSIGN	.300E+01	.299E+00			
				.600E+01	.000E+00			
				.299E+00	.000E+00			
				.000E+00	.000E+00			
						2	.015	
.319E+00			FREE	.300E+01	.299E+00			
				.200E+01	.100E+01			
				.299E+00	.000E+00			
				.000E+00	.000E+00			
						0	NO RELEASE	TRM1
						0	NO RELEASE	TRM2
						0	NO RELEASE	TRM3
						4	.000	MACH
		MACH	AWAIT	.300E+01	.299E+00			
				.200E+01	.100E+01			
				.299E+00	.000E+00			
				.000E+00	.000E+00			
			ASSIGN	.300E+01	.299E+00			
				.200E+01	.100E+01			
				.299E+00	.000E+00			
				.000E+00	.000E+00			
						5	1.608	
.548E+00		C1	CREATE	.000E+00	.548E+00			
				.000E+00	.000E+00			
				.000E+00	.000E+00			
				.000E+00	.000E+00			
		W6	AWAIT	.000E+00	.548E+00			
				.000E+00	.000E+00			
				.000E+00	.000E+00			
				.000E+00	.000E+00			
						0	.000	T1
		T1	ASSIGN	.000E+00	.548E+00			
				.000E+00	.000E+00			
				.000E+00	.000E+00			
				.000E+00	.000E+00			

Appendix B (continued)

SLAM II TRACE BEGINNING AT TNOW = .0000E+00

		NODE ARRIVAL				ACTIVITY SUMMARY		
TNOW	JEVNT	LABEL	TYPE	CUR ATRIB BUFFER		IND	DURATION	END ND
						0	.000	TR
		TR	AWAIT	.100E+01	.548E+00			
				.600E+01	.000E+00			
				.548E+00	.000E+00			
				.000E+00	.000E+00			
						1	.015	
.563E+00			COLCT	.100E+01	.548E+00			
				.600E+01	.000E+00			
				.548E+00	.000E+00			
				.000E+00	.000E+00			
			ASSIGN	.100E+01	.548E+00			
				.600E+01	.000E+00			
				.548E+00	.000E+00			
				.000E+00	.000E+00			
			FREE	.100E+01	.548E+00			
				.600E+01	.000E+00			
				.548E+00	.000E+00			
				.000E+00	.000E+00			
			ASSIGN	.100E+01	.548E+00			
				.600E+01	.000E+00			
				.548E+00	.000E+00			
				.000E+00	.000E+00			
						2	.010	
.573E+00			FREE	.100E+01	.548E+00			
				.300E+01	.100E+01			
				.548E+00	.000E+00			
				.000E+00	.000E+00			
						0	NO RELEASE	TRM1
						0	NO RELEASE	TRM2
						0	NO RELEASE	TRM3
						0	.000	MACH

References

Adiga, S. (1989). Software modeling of manufacturing systems—A case for an object-oriented approach. *Ann. Oper. Res.* **17,** 363–378.

Ben-Arieh, D. (1986). Manufacturing system application of a knowledge based simulation. *Proc. 8th Ann. Conf. Comput. Ind. Eng.* pp. 459–463.

CACI (1983). "SIMSCRIPT II.5 Programming Language." CACI, Inc., Los Angeles, California.

Carson, J. S. (1986). Convincing users of model's validity is challenging aspect of modeler's job. *Ind. Eng.* June, **18**(6), 76–85.

Caruso, P. C., and Dessouky, M. M. (1987). Analytical factors concerning the use of micro-mini storage devices as material management systems. *In* "Proceedings of the 1987 Winter Simulation Conference" (A. Thesen, H. Grant, and W. Kelton, ed.) Society for Computer Simulation, San Diego.

Chang, Y. L., Sullivan, R. S., Bagchi, U., and Wilson, J. R. (1985). Experimental investigation of real-time scheduling in flexible manufacturing systems. *Ann. Oper. Res.* **3,** 355–377.

Falkner, C. H., and Garlid, S. (1987). Simulation for justification of an FMS. *In* "Simulation," pp. 75–84. IE & M Press.

Godziela, R. (1986). Simulation of a flexible manufacturing cell. *In* Proceedings of the 1986 Winter Simulation Conference," (J. Wilson, J. Henriksen, and S. Roberts, eds.) pp. 621–627.

Gordon, G. (1975). "The Applications of GPSS V to Discrete Systems Simulation." Prentice-Hall, Englewood Cliffs, New Jersey.

Grant, H. F. (1987). Simulation and factory control—An overview. *Proc. Int. Conf. on Prod. Res. 9th,* Manitoba, Canada, pp. 576–583.

Groover, M. P. (1980). "Automation, Production Systems, and Computer-Aided Manufacturing." Prentice-Hall, Englewood Cliffs, New Jersey.

Haider, S. W., and Banks, J. (1986). Simulation software products for analyzing manufacturing systems. *Ind. Eng.* July, pp. 98–103.

Jones, C. V., and Maxwell, W. L. (1986). A system for manufacturing scheduling with interactive computer graphics. *IIE Trans.* pp. 298–303.

Klein, B. (1986). SIMFACTORY tutorial. *In* "Proceedings of the 1986 Winter Simulation Conference," (J. Wilson, J. Henriksen, and S. Roberts, eds.), pp. 530–542. Society for Computer Simulation, San Diego.

Law, A. M. (1986). Introduction to simulation: A powerful tool for analyzing complex manufacturing systems. *Ind. Eng.* **187**(5), 46–63.

Medeiros, D. J. (1989). Simulation applied to manufacturing systems. "Economic and Multiattribute Evaluation of Advanced Manufacturing Systems," pp. 384–407. Prentice-Hall, Englewood Cliffs, New Jersey.

Olumolade, M., Norrie, D. H., and Fauvel, O. R. (1990). Object-oriented integration and simulation of manufacturing. *In* "Proceedings of the 1990 Summer Computer Simulation Conference," (B. Svrcek and J. McRae, eds.), pp. 249–252. Society for Computer Simulations, San Diego.

Pegden, C. (1985). "Introduction to SIMAN." Systems Modelling Corporation, Philadelphia, Pennsylvania.

Pritsker, A. A. B. (1987). "Introduction to Simulation and SLAM II." J Wiley, New York.

Roberts, S. D. (1983). "Insight Users Manual." SysTech Inc., Indianapolis, Indiana.

Sayers, R. D. (1989). Simulation modeling at TRW provides insight on assembly operations. *Ind. Eng.* August, **21**(8), 20–24.

Shannon, R. E., Mayer, R., and Adelsberger, H. H. (1985). Expert systems and simulation. *Simulation* June, pp. 306–310.

Vig, M., Dooley, K., and Starr, P. (1989). Simulating cell activities in the CIM environment. *Manuf. Eng.* January, **101**(1), 65–68.

Wortman, D. B., and Miner, R. J. (1986). Designing flexible manufacturing systems using simulation. *Proc. Autofact* **6,** 24-29–24-55.

CHAPTER 5

Databases in Flexible Manufacturing Systems

William D. Engelke
Research Institute
University of Alabama in Huntsville
Huntsville, Alabama

I. Introduction

A. THE MEANING OF A DATABASE IN FLEXIBLE MANUFACTURING SYSTEMS

A database is a necessary and integral part of any flexible manufacturing system (FMS). The most basic function of a database in an FMS is providing storage and availability of information needed to control the manufacturing pro-

HANDBOOK OF FLEXIBLE MANUFACTURING SYSTEMS

cess. In performing this role, the database handles information needed for essentially all tasks of controlling the FMS. These include planning machine loading and management of the software used to control programmable manufacturing equipment such as numerically controlled (NC) machines. Also, databases function to interface the FMS with other computer-centered manufacturing systems such as material requirements planning (MRP II), computer-aided design (CAD), computer-aided manufacturing (CAM), and inventory control systems. See Figure 1.

Databases operate in conjunction with software (computer programs) that carry out movement and organization of data, combined with necessary computations and decision-making processes. It has been estimated that 15 to 30% of the cost of a flexible manufacturing system is incurred in control software (Greenwood, 1988a), with 50 to 80% of that being devoted to contingency management (Greenwood, 1988b) (handling exceptions, special cases, and error recovery). Databases and software are designed together and are inseparable. For a successful FMS implementation, good database design is a must.

First, it is important to understand what a database is. It is much more than simply a large computer file. A *file* is a collection of data. It may be organized so that it is retrievable only in the order in which it was stored (sequential) or so that it can be retrieved in any order desired (random). In general, a file contains just one type of data (e.g., a list of items in inventory) and all records are laid out in the same way. A database differs in many respects, because a database

- Is likely to contain many files. It is much more than simply a computerized filing cabinet.
- Has a logical structure: The topology of the database is optimized to the applications which use it. The database system for an FMS is a tightly fitting overlay on the manufacturing business it serves. Clearly, this means that it is necessary to understand the business quite well to properly design a database to serve it.
- Works in conjunction with application software to create a manufacturing data system supporting the factory operation.
- Is frequently designed to be able to serve many applications concurrently, in a coordinated fashion; the FMS may share a database with other parts of the business.
- Must be considered a strategic business resource, much as plants and equipment. It requires maintenance, staffing, occasional replacement of obsolete facilities, and can be used to generate revenue.
- Can be used for diverse purposes, including quality control, productivity measurement, process planning, optimization of machine utilization, simulation, and many others.

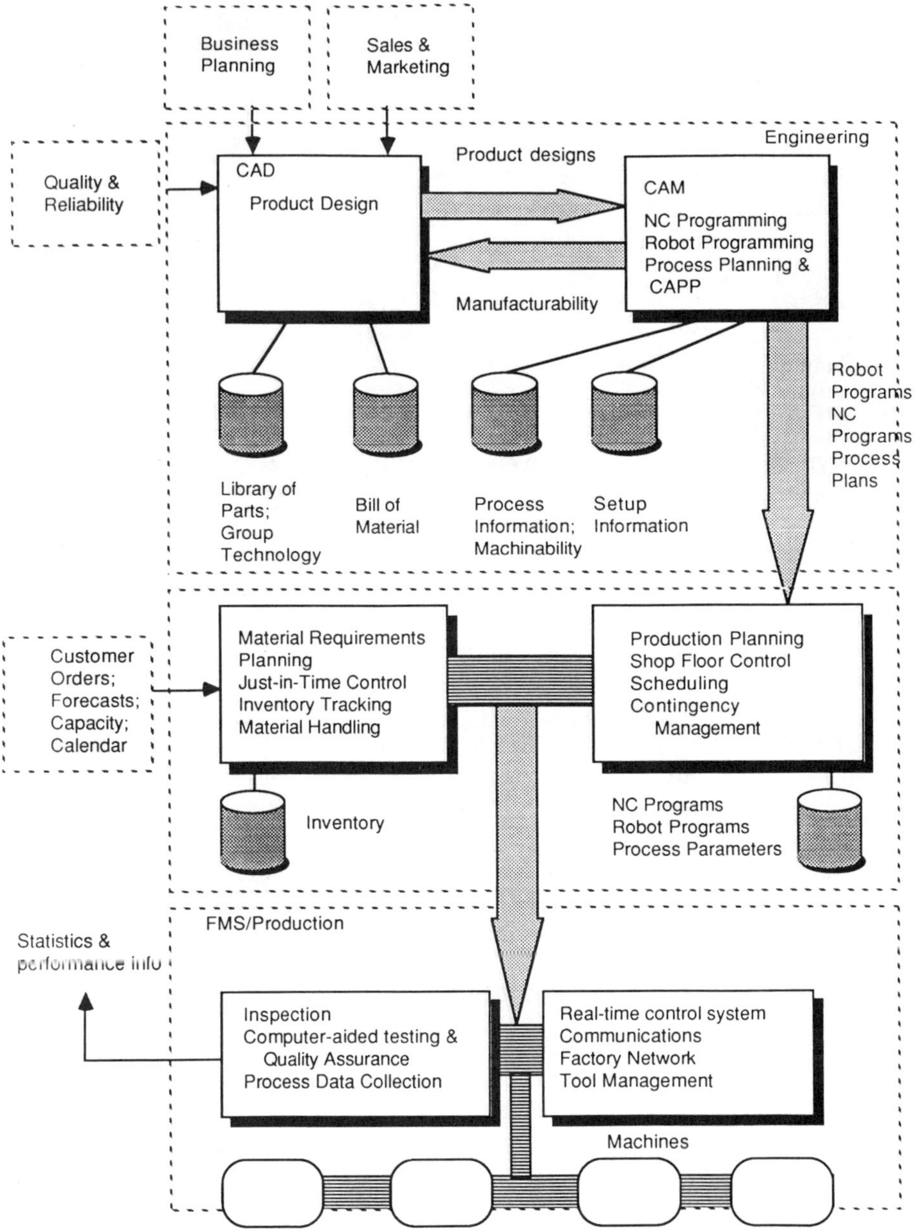

Figure 1. Typical system design and corresponding database layout for flexible manufacturing system.

Actual functions of planning and NC part programming are performed by software which uses the database as a source and repository of information. Databases can be designed so as to greatly facilitate the operation of software (or retard it if improperly designed).

Since each FMS database is tailored closely to the requirements of the individual business, and each business is different, it follows that each FMS database will be different. There is no one "best" database layout or architecture. This chapter is devoted to exploring the issues a system developer must face when designing databases to serve an FMS. Several database layouts are explained to illustrate the underlying principles behind FMS database design. Interfaces between FMS databases and other management information system (MIS) and manufacturing databases are explained.

B. IMPORTANCE OF A DATABASE IN THE FLEXIBLE MANUFACTURING SYSTEM ENVIRONMENT

An FMS database is a key element of a strategy aimed at ensuring integrity and reliability of information in the FMS system. A database management system (DBMS) is a collection of programs which provide coordinated access to information stored in the database. As part of implementing a DBMS, companies often establish a data dictionary to create a central reference which defines the content and format of each data element. A DBMS's most important function is to integrate the data so that logical relationships necessary to operation of the business are maintained. This can reduce the storing of redundant pieces of data. Also, most DBMSs provide independence of database record format from application programs. This reduces the need for massive revisions of existing programs as the database is revised to comprehend changing business needs.

C. DATABASE TECHNOLOGIES

Databases are optimized to the requirements of the applications for which they will be used: function dictates form. This represents trade-offs between software complexity, required computer capacity, storage space, ease of use, and speed of access. Commonly used database structures are *flat, hierarchical, relational,* and *network*. Frequently, for a given application, it would be possible to implement the required database using any of these strategies. To decide on the best database method for a specific case, it is necessary to understand the strengths and weaknesses of each of the methods and compare them to the needs of the project.

1. Data Relationships

Objects that we wish to keep track of in a database system can be related in several ways:

- *One-to-one*. Here, there is a direct correspondence between information items. This is often used to relate data items which are both unique. For example, if we have a warehouse with individually numbered storage bins, each bin can hold only one item, and each item has a unique serial number, there would be a one-to-one relationship between warehouse bin and item serial number.
- *One-to-many*. In this relationship, one item is related to several. For example, consider the machines in a plant. The plant contains many machines, but each machine can be located in only one plant.
- *Many-to-many*. Here, an item can be related to several others, but the relationship goes in both directions. This is the case in the product-machine relationship in an FMS. A specific product must be processed on several machines, and any specific machine is capable of processing several different products.

Methods for handling these relationships differ between database methodologies. A workable FMS database design is likely to contain all three of these types of relationships, and any practical DBMS must provide a method of coping with them.

2. Flat Databases

A flat database does not have a logical structure apart from its basically sequential nature. Depending on the level of sophistication of the computer system, a flat database may allow random access, and sequential access in different sequences via secondary indices. For example, it may be possible to sequentially access all the parts in a component database by part number (primary index), by date purchased (secondary index), or by manufacturer (also a secondary index). Flat databases are used when all the information items in the database have the same rank; none is superior or subordinate to another, or used in another. This would be the case with databases such as a list of single components, a series of orders for the same type of item, or a list of customers. The one-to-one relationship is well suited to representation via a flat database.

Theoretically, it would be possible to keep all the data for any given application in a flat database. In a case like this, each record of the database would have space for all possible information items (fields); this is termed a *composite record* (see Figure 2). In this example, a database contains all a factory's machines and all the parts they can manufacture. Note that since a particular machine can manufacture several different parts, that machine is listed several times in the database, with each of the different parts it can build.

Plant	Department	Machine	Part	Engineering Change Level		etc.
12	101	223	54432	C		
12	101	223	57665	A		
12	101	223	54432			
12	101	223	65567	F		
12	101	223	67832	AA		
12	103	224	54421	G		
12	103	224	55643			

•
•
•
•

Figure 2. Composite record layout in a "flat" database: All fields of information for all applications are included in each record.

A database built of composite record often contains a vast amount of data redundancy, that is, data items stored repetitively. In this example, even though there is only one machine No. 223 in the factory, it is listed in the database five times because it can manufacture five different parts. The impact of this redundancy is increased space requirements to store the data, and longer search times to retrieve the data when needed. This situation grows geometrically with increasing complexity, making it impractical to handle most real-world applications using just flat databases. In the following sections, structured databases which address this problem are discussed.

3. Hierarchical Databases

It frequently happens that information items in the real world have a hierarchical (or tree) structure, in that items are related to each other by a superior-to-subordinate relationship. In database terminology this is often termed a "parent-child" relationship. A good example is a bill of material (BOM). See Figure

3, a *structure diagram* (SD) for a hierarchical BOM database. An assembly is known by a single part number. This assembly is made up of several subassemblies, each of which has a part number. This breakdown continues until it reaches an "atomic" level at the individual component level (or, a BOM structure may even go all the way down to bulk materials used in the manufacture of the components). What is notable about a BOM structure is that all different levels

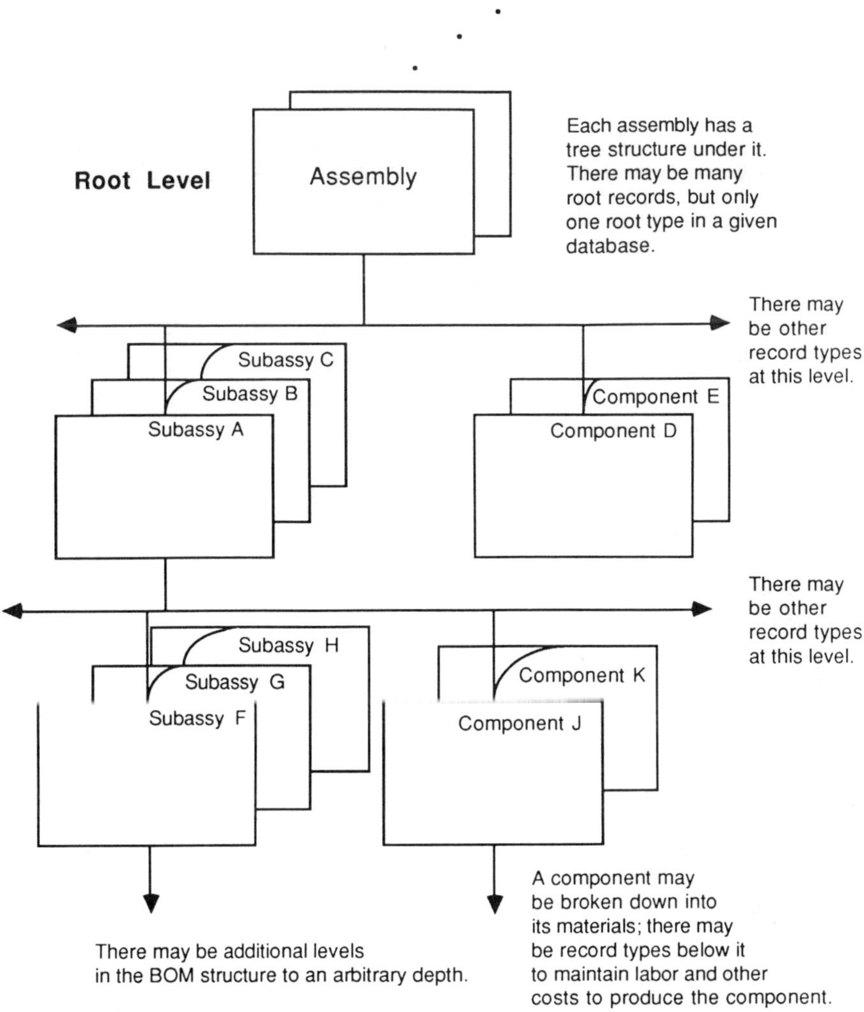

Figure 3. Structure of a hierarchical database. A bill of material may contain many levels, with each assembly (major top-level part number in this case) having a single root-level record.

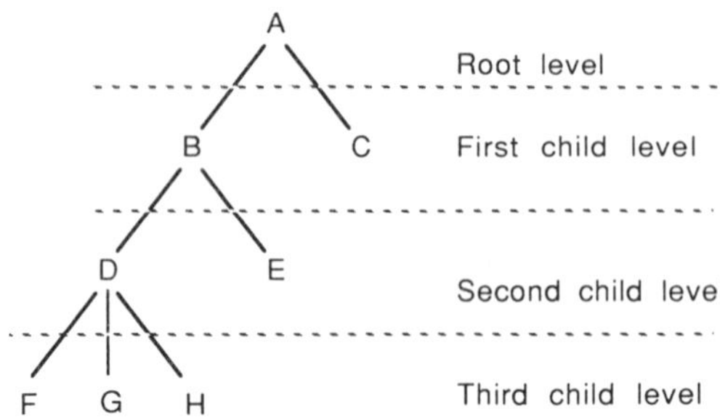

Figure 4. Structure diagram (SD) showing a typical hierarchical database layout. The various records shown (A, B, C, etc.) may be the same or different record types.

of the hierarchy of the assembly are stored in the same database. The uppermost level of the hierarchy, in this case assemblies, is known as the *root level* of the database. It is possible to access the database at the root level to simply list all available assemblies. The database may also be traversed at any lower level. It is also possible to navigate the database by moving up and down the parent–child relationships between the levels. In this way, it is possible to retrieve all the components used in an assembly (starting from the top level), or determine on what assembly a component is used (starting from the bottom). This type of structure is very widely used because it parallels the actual relationships between real objects quite well. Many such relationships are found in a manufacturing environment, including information oriented toward product, process, tools, machines, conveyor lines, orders, and so forth.

The SD which describes a hierarchical database shows the parent-child relationship between information items. Note that the vertices of the SD do not represent individual occurrences of data items, but types of data items (Figure 4). Here, a database contains a single root (by definition) type A; subordinate to the root are item types B and C; subordinate to B are D and E; and so forth. An item type can have multiple occurrences. Underneath each occurrence of that item type, there may be multiple occurrences of the item subordinate to it. Hierarchical databases are well suited to representing the one-to-many relationship.

4. Relational Databases

There are some real-world relationships which are not easily represented by a hierarchical database. The hierarchical paradigm breaks down when it is necessary to create a structure where a "child" record has multiple "parents." This is the case in the many-to-many relationship. It is possible to force this structure upon a hierarchical database, but doing so will require much redundancy of information and possibly also exact a severe performance penalty. Note here that redundancy in and of itself is not necessarily a problem. In the BOM example just cited, consider the following requirement. If a particular component is used in multiple assemblies, it will appear in multiple places in the database. If we must do a "where-used" search (in which we start out with a given component and search for all assemblies in which it is used), a lot of searching will be required in a strictly hierarchical schema.

The database architecture known as relational can address this situation. In a sense, a relational database is a series of flat databases (each on one topic) linked together by relationships. A composite record database, as described earlier, can be split into several files called tables to form the beginnings of a relational database.

Consider the example of a factory's machines and parts (Figure 5). The composite record file would be broken into two files, one containing a list of all the parts, and one containing a list of all the machines. This split has lost some intelligence; it must be restored by creating a relationship file that links machines and parts.

In the case of the BOM, the relationship file can serve to point from assemblies to subassemblies to components, down through the product hierarchy. In that case, the relationship file contains relationships between records in a single file. A where-used relationship can also be inferred from this same file,

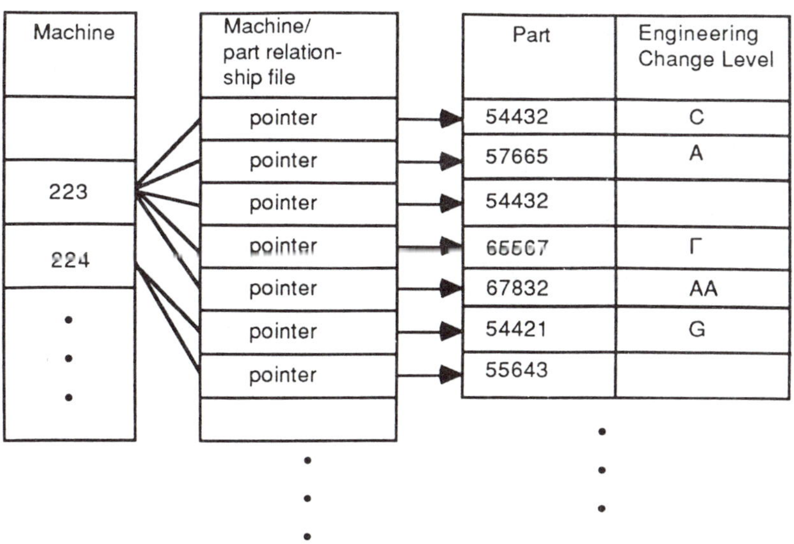

Figure 5. A relational database reduces data redundancy by storing most fields of information only once. Relationships are maintained by relationship files which point back and forth to related information. These are sometimes called "intersection" records.

but for performance reasons it is likely that it would be implemented as a separate relationship file with a different sort sequence. Even this method may not prove entirely satisfactory in this many-to-many relationship. Refer to the following section.

5. Network Database

The network database architecture is well suited to handling the many-to-many relationship. It can represent the many-to-many situation with a good trade-off between redundancy and performance. Network databases are also good for handling complex, tangled relationships that do not lend themselves well to hierarchical representation (e.g., the transitive graph). This situation can occur in FMSs that handle a number of product lines of varying complexity, especially when subassemblies are used at different product levels depending on the order, customer, and so forth.

Groups of related data items are known as *sets,* and the data items are members of the set (see Figure 6). (This does not follow all the same rules as mathematical set theory.) A set is created by linking related items in a circular loop which starts with the set owner, travels through each member, and ends at the owner. In the example of the factory machines and parts, a set begins with a machine, traverses through the parts it can manufacture, and ends at the machine. An item can be a member of more than one set. This would be a frequent occurrence in this example, as a typical part requires more than one

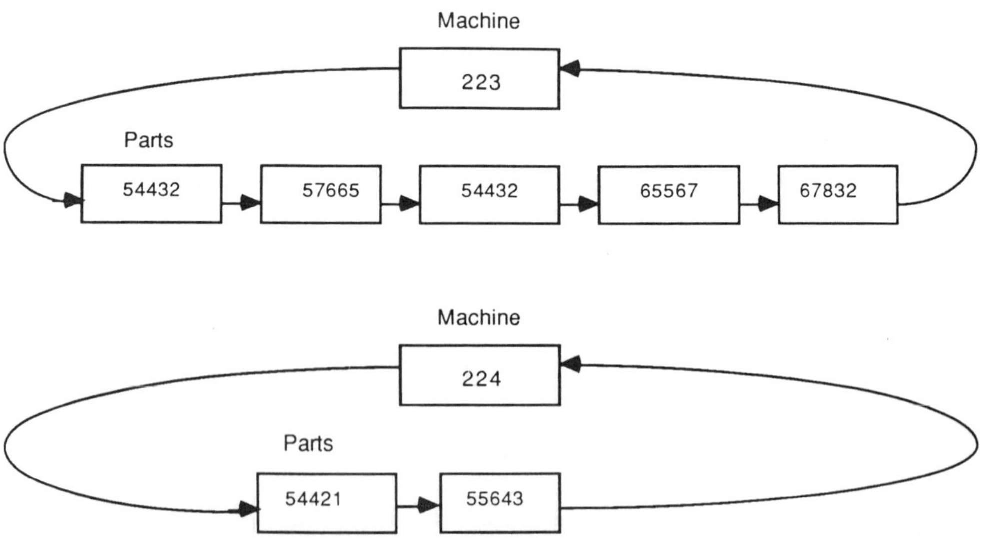

Figure 6. A network database stores related information in sets.

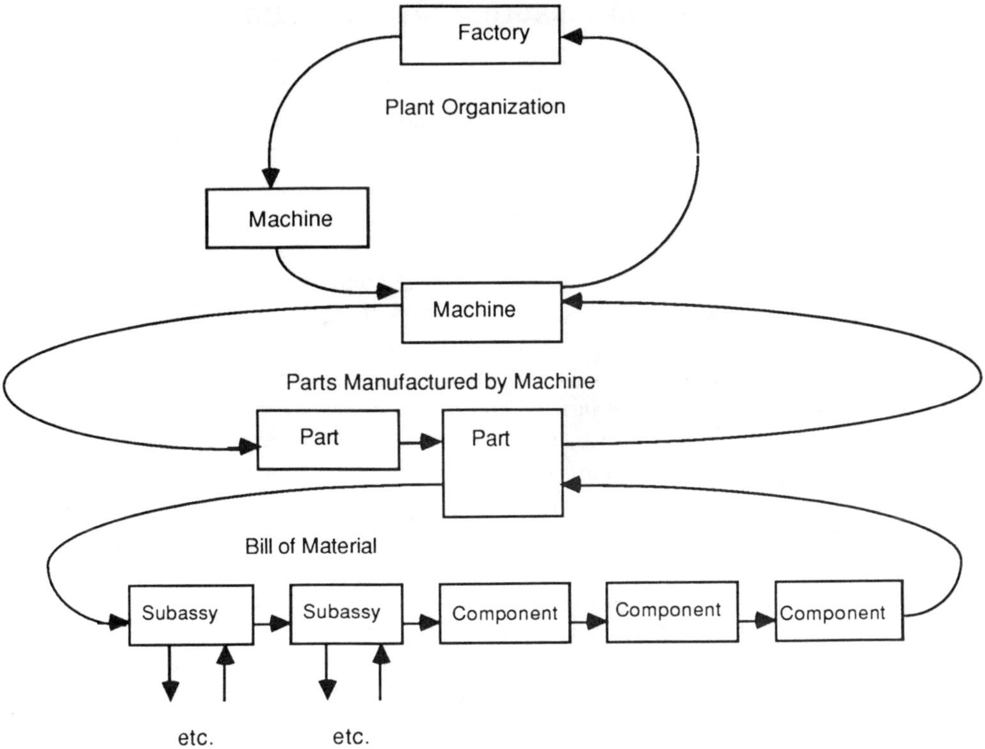

Figure 7. A record in a network database may be a member of multiple sets.

machine to build it. Also, the machines are part of another set, the factory set, which contains all machines in the factory. As the network set SD Figure 7 shows, the factory, machine, part, and BOM sets can all be related. The many-to-many part-machine relationship is implemented via a *link* which acts as a bidirectional intersection between the sets.

All the database architectures just discussed are implemented on computers using a variety of technical methods, including random access files, indexed files, hash tables, binary search trees, pointers, secondary indices, and many other techniques. It is important that the team which is going to be selecting database technologies be familiar with these technical methods to avoid selecting architectures which will later prove unworkable after much investment has already been made. The primary risk factor in designing databases for FMS is performance. The wrong database architecture and/or design can reduce performance enough that manufacturing schedules will be impacted. This will then lead to a requirement for expensive computer upgrades that will hurt the economic viability of the project.

II. Manufacturing Data Systems

A manufacturing data system (MDS) acts as the link between the abstract, computerized worlds of sales, engineering, and production control, and the real world where actual products are built. The MDS carries out the diverse functions of computer-aided process planning (CAPP), scheduling, inventory control, and manufacturing requirements planning (MRP). The FMS database is part of the interface between the MDS and the factory floor.

Figure 1 shows the layout of a typical MDS and how it fits into the overall MIS used to operate the business. Note the flows of information between various subsystems. Individual companies will have more or less computerization of information flows depending on their situation. (Sometimes a database on paper or a verbal data flow works just as well as the computer version). These data flows ultimately direct manufacturing to produce the desired quantities and product mix. Sophisticated systems will also incorporate capabilities to optimize production. This may include such subsystems as machine loading optimization, automatic routing of work-in-process, and even product mix control to maximize profit.

Generally, each subsystem of any MIS can act as a data source and as a data receiver. The MDS receives orders from other parts of the MIS regarding what products should be produced. The MDS can also be a rich source of data on the manufacturing process. With proper design, the MDS can be structured to provide management and manufacturing engineering with valuable feedback on production rates, machine downtime, plant efficiency, inventory, and many other plant conditions.

The MDS forms the environment in which the FMS database operates. The FMS database cannot be considered in isolation because it depends on the MDS for input.

III. Manufacturing Data Flow

A. OVERVIEW

Refer again to Figure 1, noting the overall flow of information in a manufacturing operation. Note how the FMS acts as a data receiver primarily, but also as a data source. In a typical system, these databases and software modules would be distributed across several machines, with data flows traveling across communication lines. Section VI discusses distribution of data and functionality through a system hierarchy.

In the FMS, automation of data flow is as important as automation of material flow. While manufacturing is being performed, the software which controls the operation requires rapid access to information to make decisions. A necessary part of FMS database design, therefore, is a detailed articulation of the procedures which will ensure data availability and integrity. For example, a master production scheduling program needs information on customer orders and/or inventory stocking requirements, material availability, and capacity (including machine/cell availability) to develop a manufacturing schedule. Even more information is required if the scheduler is to perform any optimization. Manufacturing data systems often consist of a mixture of batch and real-time processing operations. Clearly, it is necessary to lay out a plan for synchronizing certain processing activities which load the pertinent databases. This also must include contingency plans to cope with computer downtime.

B. UPSTREAM–DOWNSTREAM ISSUES

From the notion that FMS is primarily a data recipient it follows that somewhere data must be created. A number of automated systems provide these data, but also human-driven systems are affected. The CAD area is one in which changes from traditional operating procedures are needed to make FMS effective.

In the past, CAD groups were often operated basically as electronic drafting shops. There is a growing awareness, however, that CAD procedures and the responsibilities of upstream engineering sections need to change significantly if full advantage is to be taken of FMS capabilities. Specific implications of this include

- The product and the process to build it must be designed concurrently, by a closely knit team.
- CAD groups must often do some extra work that they did not do before so that their data will be useful to CAM and FMS. This means absorbing new costs and leadtimes.
- Product design groups need to develop a much-increased understanding of design issues related to producibility, quality, and reliability.
- Reorganization of engineering activities should be considered at the executive staff level so that the above points can be considered in the management reward structure.

A specific case that illustrates this principle is the issue of three-dimensional (3D) design. 3D design takes more work and may be considerably more difficult than two-dimensional (2D) drafting (paper or electronic). However, a 3D design contains considerably more intelligence than a 2D drawing. It will often be the case that 3D design needs to replace 2D in CAD shops with the advent of FMS. In a typical large CAD/CAM system implemented by a division of General Motors, it was found that the engineering design group's costs were 46% of the total cost of the system, but that group gained only 21% of the benefits.

Conversely, the tool room group bore only 20% of the cost of the system but reaped 45% of the benefits (Engelke, 1987). Clearly, this system could enjoy support only as a team effort including the various involved staff areas.

Engineering change tracking can be a major function of the FMS database. This includes control of the design being produced, future and past versions of it, manufacturing data for various change levels, and may extend to control of inventory to ensure that usage of the components specified for each engineering design version are coordinated. When automated manufacturing processes are being used, especially with low production volumes, the integrity of the design is vital. It is important to avoid having multiple copies of the design, containing inconsistent changes, circulating. An early decision to prohibit manual changes to designs and specify that the CAD design is the master can enforce the necessary discipline.

IV. Computer-Aided Design Considerations When Planning for Flexible Manufacturing Systems

Computer-aided design (CAD) is the point where design intent takes on electronic form, and hence can facilitate or impede the manufacturing operation. Often CAD shops which have been in operation for many years are forced to rethink their approach to design and drafting to enable the company to enjoy the advantages of flexible manufacturing. The following are tools which can help with the implementation of this direction.

A. LIBRARY OF PARTS

A library of parts is a database which aids the product designer who is faced with a very large selection of potential components when creating a design. Its major application is as a catalog of parts to be included in an assembly. In many cases, it is necessary to custom-develop a library-of-parts system optimized to the product, business, and company to be using it.

To be a powerful design aid, the library of parts requires the following capabilities (see Figure 8):

• It must be searchable so that designers can let the computer do the work of selecting the right component for a specific application from a large catalog (perhaps tens of thousands).

• It must provide ancillary information on parts that meet search criteria so that the designers can decide if parts suggested by the computer will meet the need. This may be a display of a graphical representation of the part, a list of key dimensions or component values, or other information.

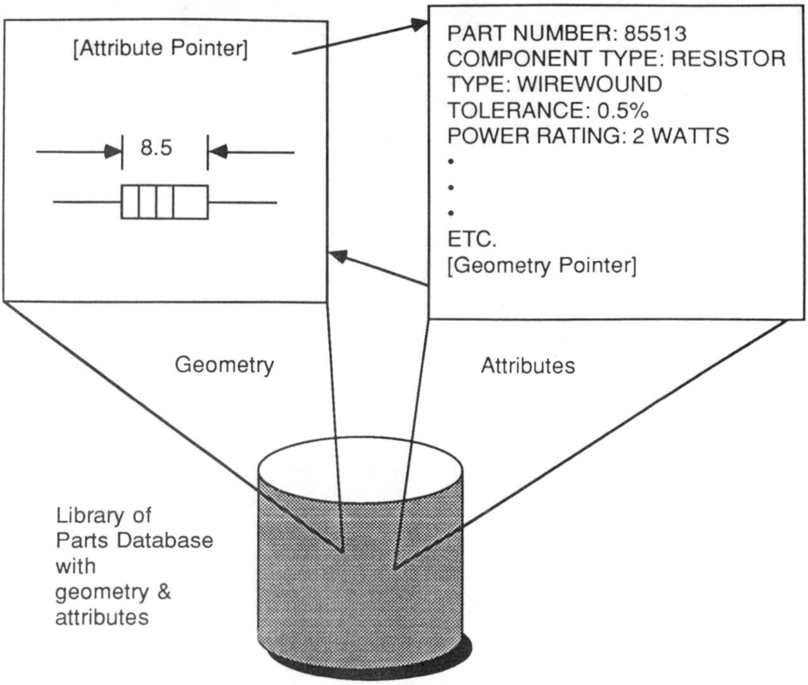

Figure 8. Attributes are related to geometry logically. These logical relationships are embodied in database architecture.

Additionally, a number of other functions may be required, such as an expert system which assists less experienced designers by using a knowledge base developed by seasoned experts, automatic design optimization for cost/performance, and reference to previous design detail, to name a few. The product designer must be involved in the design of the database and search system to specify search and display capabilities.

When a fully functional library of parts is available, it is often found that much more efficient use of available parts can be made. For example, if it is easy to search the computer to see if a certain type of part already exists, there will be fewer new designs specified which largely duplicate existing parts. With proper design, the library of parts can also be designed to control the use of obsolete parts, control engineering changes, manage supplier usage based on quality, and forecast parts usage and final assembly costs.

Ancillary information about each component is generally stored in the library of parts along with the design so that it can be included in an assembly. When component attributes are stored in the assembly, they may be extracted by software to produce files required for assembly, machining operations, bills of material, and other uses (see Section C.)

Needless to say, all of these undeniably useful functions require an investment. Designing, populating, and maintaining a library-of-parts database can be very labor-intensive and time-consuming. For example, a division of General Motors found that, when creating a library of electrical parts with approximately 25,000 components, it was necessary to expend one-half to one man-hour for each part added to the system. (This database contained 20 fields of information and a drawing on each part.) After initial creation, three to five changes per day were made to reflect newly available parts and engineering changes. It was calculated, however, that tooling costs to create a new part were such that even if the library of parts reduced the incidence of new parts by only 5 to 10%, the library would be cost-effective.

B. GROUP TECHNOLOGY

Group technology (GT) is a fairly broad term which refers to the practice of taking advantage of similarities between different parts to enhance the design and manufacturing operation. Parts may be grouped by shape, machining/processing requirements, material, or any other convenient attribute; GT may be used in conjunction with a library of parts, if desired. Some manufacturing operations do not need a sophisticated library of parts because they are building parts only, not assemblies. In that case, a simpler library using GT can be used for process planning, optimization of machine loading, or checking for the existence of a similar part before tooling up to build a new one.

Group technology is not a database methodology as such. A GT system must be set up in accordance with the needs of the product and the business and can then be adapted to fit in with the MDS. As a basis of storage-and-search techniques for a library of components, GT can make the database more understandable and accessible to people who have little computer training.

C. SPECIAL COMPUTER-AIDED DESIGN REQUIREMENTS FOR FLEXIBLE MANUFACTURING SYSTEM DATA PREPARATION

When product and process are designed together, a product model can be developed which contains all the information necessary for effective interfacing to the FMS. The usefulness of a model created in CAD is directly proportional to its intelligence content. The two major facets of intelligence are *geometry* and *attributes*.

1. About Geometry

The geometry of a part is the computer's model of its physical shape. Typically, the minimum requirement in a CAD model destined to be used for FMS is a 3D wire frame. This level of detail captures front-to-back, side-to-side, and relative-fit relationships. It can be of use for FMS if parts are straightforward and not easily misinterpreted; this implies that CAM engineers developing tool paths can easily understand the nature of the required surfaces. When part designs may contain ambiguities in a wire-frame model, it is necessary to add surface representation to the part. This is also the case when sculptured surfaces are required. Solid modeling offers the ultimate in part intelligence content, but adds little that is of use to the CAM engineer.

As part of the effort to integrate product and process design, the creation of CAD parts must consider the needs of FMS. Specific needs depend on the system and product. Typical examples:

• Some CAM packages require that the end points of lines have *identical* coordinates to be considered joined. (This is necessary for the system to generate instructions that will, say, allow a cutting tool to continue around a corner.) However, it is easy to inadvertently create a part design where the end points look as if they are joined but differ by a minute amount too small to be visible on the screen. The discrepancy will later create annoying fix-up work for the CAM engineer.

• It is important that design intent information in the CAD model be complete enough so that the CAM engineer can create tool pathing and other required planning by reference to the design. Some of the advantages of using CAD/CAM are negated if the CAM group must constantly be on the phone to the CAD department while preparing FMS data. This may require the creation of software customized to the product so that all required questions are answered up front.

2. About Attributes

Attributes are information items about a part which are stored along with the CAD design; they contain data that may or may not be displayed when the part is viewed or plotted. These attributes can be created manually when the CAD designer draws the part or created by engineering-assistance software which queries the designer for information (see the following section).

Once a design has been created, software can be created which extracts attribute information as required for use in various FMS subsystems.

Design of the attribute database for a product is based on the product's characteristics and the needs of downstream applications. For an assembly, a database often will have a hierarchical structure. This allows reference to subassemblies at any level. In some cases, database topology may reflect technical considerations, such as the need to analyze the load on a structure or current in a branch of a circuit. The database in this situation acts as a model of the product. The database may also be (topologically) very simple in the case of machined components.

3. Design Cycle Data Preparation

To take full advantage of the capabilities offered by a joint geometry/attribute database, custom software must be developed to aid the designer (Figure 9). Here, a library of parts with attributes has been created. Also, custom software has been written which interacts with the designer during the design process. This has several functions:

- To ensure that engineering design standards are followed
- To audit the design to prevent certain (predictable) errors and omissions from occurring
- To do custom computations unique to the product (e.g., calculating precise quantities of required materials even in a nonscaled drawing, checking certain technical attributes such as current, flow rate, etc.)
- To handle certain repetitious functions
- To recommend or automatically select features based on required parameters
- To force the designer to pay attention to manufacturing constraints such as process limitations, producibility, and quality concerns

Automatic decision making can be incorporated by the use of artificial intelligence; an expert system making use of the experience of many designers can be developed to interact with the designer if appropriate.

As the designer works at the CAD station, a geometry/attribute database is built which is specific to this particular part or product. After completion of the design process, another custom program is run which extracts information from the product's attribute database suitable for use in other MDS functions. Depending on requirements, CAM engineers may also make use of the 3D model to develop NC programs and/or robot software. Completed manufacturing data then are moved to databases which are accessible during the scheduling and manufacturing operation.

Another important part of design data planning is standardization of drawing/model content. Since models are used by multiple departments in an integrated system, standards must be set and adhered to for entity types, geometric presentation, level content, scaling, surface representation, completeness of model data, and possibly others. Custom application software can include capabilities to create designs that follow the standards.

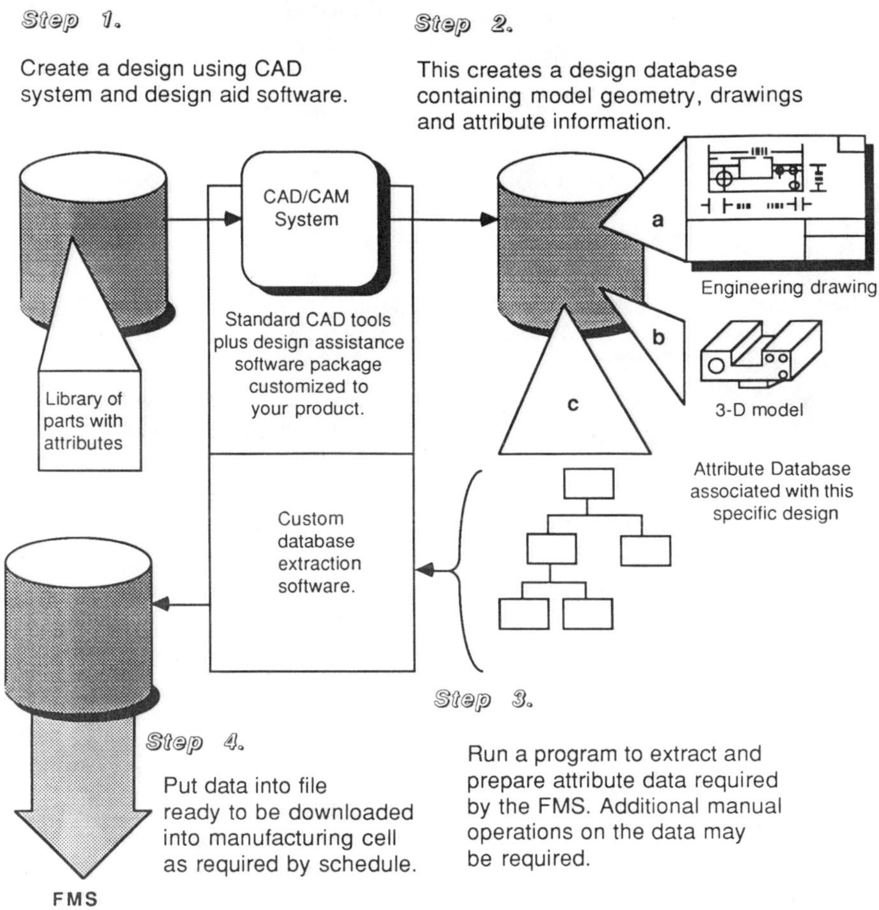

Figure 9. Steps in going from CAD to CAM to FMS ready data.

V. Computer-Aided Manufacturing Considerations When Planing for Flexible Manufacturing Systems

Adding a new part to the FMS is a multistep procedure which includes consideration of requirements for machining, raw material, material handling, and inspection. Each additional item then becomes part of the repertoire of parts that can be produced in the FMS. Manufacturing data therefore require the same level of control as is provided to tracking engineering changes and design levels in the CAD arena.

Considerable care must be given to planning the flow of data from CAD to CAM to numerically controlled (NC) machines. Assuming CAD data are complete, it is still necessary to do quite a bit of work to prepare a tool path. Along with guiding the CAM software in selecting machining paths, the engineer must set speeds, tools, coolant, and numerous other parameters. Completed NC programs must be stored in a database such that the MDS can access them in the correct sequence and pass them to the appropriate manufacturing cell as required. Part NC programming must be performed well enough in advance so that test pieces may be machined. Completed and tested NC programs can then become part of the NC database.

A typical FMS cell will have a fairly limited range of products it can build, so it is possible for NC programmers to develop a high skill level within this narrow range. Also, expert systems can be combined with machinability databases to increase the effectiveness of the NC programming operation.

Robot programming can be improved by using a simulator package for off-line programming. A number of systems are available which provide the ability to develop robot programs off-line and visualize the robot moving through its steps. A difficulty often encountered is that there are few standards in robot languages. It then becomes an entire area of database storage and control to manage the creation of the robot programs on the simulator and translation into the robot's language. Typical steps in this procedure are

- Develop 3D kinematic models of the machines to be used, including robots, conveyors, parts feeders, transfer mechanisms, and all other moving equipment. (*Kinematic* implies developing rules governing how the model's parts can move.) Vendors of off-line programming systems can supply models of many popular robots, and develop custom models for specialized equipment. These models can also include simulation of other key capabilities of automatic equipment, such as input/output (I/O) ports and programmed decision making. These become part of the database associated with FMS.
- Develop 3D models of the manufacturing cell to be simulated. This refers to stationary equipment with which moving equipment could possibly collide.
- Program simulations of the cell action sequence to be implemented. This allows engineering to check for collisions and interferences, observe the timing of operations, ensure that robot destinations are reachable, check that I/O interconnections make sense, and other factors. Simulation programs are written in a language unique to the simulator.
- Translate the simulation program into the required robot language. This can be a real challenge if the simulator language does not map well into the

robot language; some manual fix-up of the robot program will probably be necessary.

• Even though most program problems will have been cleared from robot software prior to use (by way of the simulation) there will still be discrepancies. Most notable of these are destination errors, because a manufacturing work cell is usually built with slight errors. The program and robot must now be calibrated against the actual work cell.

Preparing robot programs by simulation reduces the amount of time the expensive robot must be tied up for engineering debugging, as well as reducing the danger to personnel and fixturing due to testing new robot software. It also simplifies the creation of high-precision robot activities.

VI. The Flexible Manufacturing System Database System

The typical FMS will require a combination of databases and complementary software systems which manipulate them. This section discusses commonly needed system areas and design methods for customizing a system for a specific FMS. Additionally, it is important to plan data exchange with other MDSs and treatment of specialty areas unique to a particular business.

A. MAJOR SOFTWARE/DATABASE SUBSYSTEMS

Since software to control the FMS and FMS databases are tightly linked, they must be considered together. It will also often be found that specific databases have data interactions with multiple applications. An application may also access and update several databases. Because of this, it is vital to consider areas of required functionality rather than considering databases or software in isolation. A high level of automation in FMS control requires a fairly complete complement of "upstream" manufacturing database subsystems. Refer to Figure 10. Major software/database functions which may be required for the operation of an FMS include the following:

• *Computer-aided process planning (CAPP)/work scheduling* includes accepting order information from the MIS, planning production, scheduling machines, possibly optimizing the schedule. A machine capacity database is a vital part of this application area, as is a calendar system to allow rational scheduling of operations.

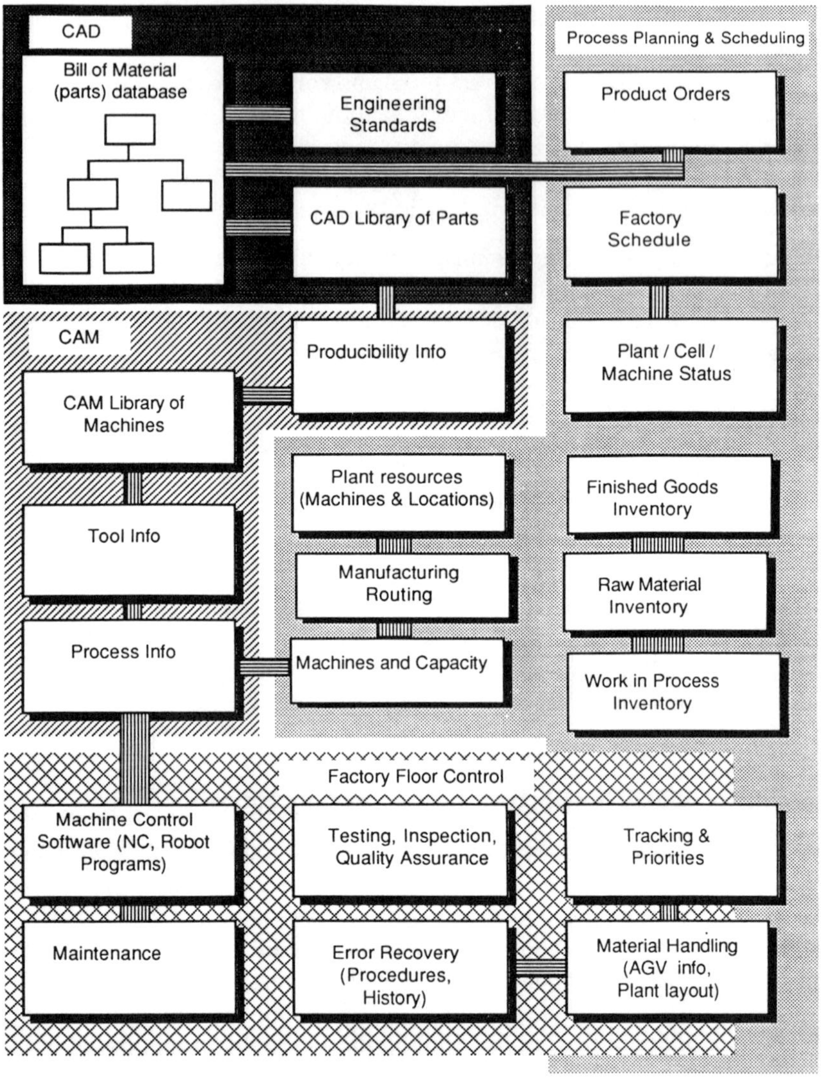

Figure 10. An example of the type of database layout that is developed to support FMSs. (Many data relationships and subsystem databases are omitted for simplicity.)

• *Component and process databases* are used by CAD and CAM groups for product and process design. These are far upstream from the FMS, but can have a big impact on FMS productivity due to the fundamental relationship between product and process design.

• *NC and robot program control* operates in conjunction with off-line programming and scheduling. It ensures that software to drive NC machines and robots is available and downloaded to the machines when required by the schedule. It may be coordinated with contingency management to assist equipment in recovering from glitches such as dropped parts. This may also include a system to make setup information readily accessible to plant personnel for machines that require manual settings to switch between parts.

• *Decision support systems* (DSSs) use data drawn from several database areas to assist in running the FMS. This may be information supplied to scheduling algorithms or to humans who are optimizing FMS operation manually. Typical areas in which DSSs are used include parts selection, batching and balancing, managing of tools, dispatching of work orders, simulation, and queueing. There may be several DSSs at different points in the hierarchy; for example, the top-level mainframe would handle dividing orders into batches (longer-term decisions), whereas an FMS-level DSS would deal with reallocation of capacity due to machine downtime.

• *Off-line programming* allows development of NC programs (tool path creation), robot programs, and often simulation to observe the actions of completed programs. Also included is a system to convert (translate) programs into the finished form required by the specific brand(s) of robot or NC machine being used. Software to perform these functions is among the most complex of the FMS operation, and numerous databases are involved, including databases for tools, materials, products, robots, NC machines, manufacturability, kinematics, and others.

• *Status monitoring* is a real-time and/or a regularly scheduled function which provides feedback on how well manufacturing is operating. This may be used primarily for keeping watch on the automated systems or may also be used for building up a historical database to analyze for downtime, mean-time-between-failures, and productivity. Communications will be involved, interrogating information stored in local programmable controllers and computers while gathering data centrally. A complete status monitoring plan must also include a plan for how these data will be used (do not collect them without a clear idea of how to make use of them).

• *Inventory* data are needed to allow planning of manufacturing based on materials on hand. Finished goods inventory may or may not be closely linked with the FMS. If automatic guided vehicles (AGVs) and automatic material handling systems are used, inventory data management is generally integrated with FMS control.

• *Material handling,* if handled by automated systems or AGVs, must be linked in real time to the FMS operation. As with contingency management, handling of exceptions will be a major part of the design. This application involves primarily control software rather than database manipulation.

• *Ancillary process data* refers to a collection of databases of highly specialized information (e.g., tool wear rates, machinability, material characteristics, speeds, and feeds) which may be used by various applications such as scheduling and optimization, CAD, and CAM. In more sophisticated systems, this type of information may be captured as rules in a knowledge base and used in conjunction with expert systems for optimization.

• *Order information* provides input to scheduling, and also forms a basis of information on which to base material requirements planning and just-in-time production control systems.

• *CAD/CAM/FMS data preparation* can be one of the most critical and complex data flow/database applications in the business. As previously discussed, initial CAD work must be done by a team which understands product producibility, the process, and data requirements for CAM/FMS. Engineering change levels must be controlled so that only the appropriate data becomes available to downstream data users. At the CAD end, libraries of parts must be maintained. The product and process will dictate requirements for CAM libraries: these may include manufacturability databases, process information, machine/robot models, and other information discussed in preceding sections. When required data preparation has been completed, provision must be made to (*a*) test NC machine and robot programs to ensure correctness before attempting to build product and (*b*) inspect the finished product to ensure adherence to specifications. This involves procedure and management control which are ancillary to the database system per se but are necessary for operation, and thus must be designed in tandem with the database system.

B. SUPPORTING DATABASE SYSTEMS

Interconnection and timing of FMS database systems, subsystems, and control require supporting activities and capabilities as follows:

• *Communications* will be a mix of networking (numerous technical approaches are available; local area networks are often used in some form, with specialized capabilities for the factory environment), direct computer-to-computer communication (e.g., RS-232 asynchronous or bisynchronous), and simple on-off levels. Local area networks (LANs) must be planned to have sufficient capability and speed, and can be a major factor in cost/performance. Interconnections between computer systems take on a major role if remote computer systems are doing part of the data preparation for the FMS, such as scheduling, creating pick lists, or handling just-in-time production control. Avoid being at the mercy of leased-line vendors to be able to run the factory.

- *Contingency management and standby operation* may form a significant part of the software effort to avoid complete loss of manufacturing capability in the event of failures. Some manual intervention will usually be required. Automated contingency management may be handled as part of a decision support system, as previously discussed (changing order size, order deletion, modify pallet contents, empty or load buffer store, etc.) or simply provide a means to notify an operator of a problem and allow some manual operation.
- *Data backup* allows the saving of database information to avoid loss of work in case of lost computer files.
- *Remote diagnostics* allow the user to test computer equipment at a distance. This is useful for tracking down the source of field failures from a central location.
- *Data exchange* with other manufacturing data systems to accept information, make inquiries, and provide information on manufacturing performance. This is a separate topic from communications in general, involving special areas of expertise. Transmittal of information between systems can take many forms, from electronic mail to automatic updating of remote, distributed databases.
- *Part programming and program verification* provides a framework for testing NC and robot programs. A large, active shop with many parts and frequent new parts and changes will also need a database system to track the development and release of these programs.

Specifications for these auxiliary supporting functions should be developed in tandem with overall general database system design, and examined for feasibility. This ensures that impossible requirements and cost structures are not built in.

C. SYSTEM DESIGN AND DATA FLOW PLANNING

1. System Engineering for a Flexible Manufacturing System Database Layout

It is possible to buy flexible manufacturing systems and many of the software packages needed to support them. Database structure, however, is best designed in-house, to ensure a good fit with the business. At a minimum, even if the FMS database system is designed by a consultant, the user should be closely involved and take an active part in developing the functional specification. The design of a database system and its associated software systems uses many of the same well-defined design techniques that are used for designing systems in general, and in particular, data processing systems, manufacturing/factory systems, and communication systems. System engineering techniques and standardized data system design methodologies are applicable. To achieve computer-integrated manufacturing (CIM), it is necessary to take a systems approach, designing the entire system as an integrated unit, including the required specialty areas, and considering the system's total life cycle. Otherwise there is a significant risk that the CAD, CAM, MDS, and FMS areas will each end up as islands of automation. Part of the systems approach also should include the development of a data dictionary which defines all data fields (including length, format, and content). This reduces duplicative effort in building systems, as the same data format is adopted companywide for each field.

An idea familiar to the system engineer is the allocation of overall system function from the uppermost levels of the hierarchy of subsystems down through the lowest. Distributed database technology combined with networking allows storage of pertinent data in the computer nearest to the application. This can help reduce the amount of data traffic required to run the manufacturing operation. Also, the increased reliability of eliminating single points of failure acts to prevent a shutdown of the entire factory in case of a computer malfunction.

The fundamental concept here is to design an integrated computer, software, and database system which meets the needs of the FMS and links to the other parts of the MDS. Usually, the optimal overall system design will be somewhere between centralization and decentralization. The proper mix of processors and communication lines will be determined by considering the trade-offs between the needs of integration (the overall system) and the needs of the individual processes. A hierarchical approach has many advantages over total centralization or decentralization. Hierarchy inherently preserves control and command over the process, necessary to carry out the management plan. Distribution of intelligence, computer processing power, and data down to the lowest possible level of the hierarchy increase the throughput of the total system by ensuring that upper-level system downtime will not immediately be felt by lower levels.

Refer to Figure 11, which illustrates allocation of database/system functionality down through a factory data processing hierarchy. Note that allocation of required database storage and software functionality often tracks closely with the allocation of manufacturing task responsibility. Individual computer systems are designed around the needs of the manufacturing functions they perform. Typically, a manufacturing cell will be coordinated by a single computer which is part of a local area network. Individual machines, transfer lines, conveyors, and ancillary equipment will be controlled by programmable controllers which are downloadable from the cell computer (or the local area network). In this scenario, functionality of the individual manufacturing machines is limited to a subset of their potential capability, but is designed to be sufficient to handle the required range of products being manufactured. It is also possible to design a system for even greater flexibility by using computers (instead of programmable

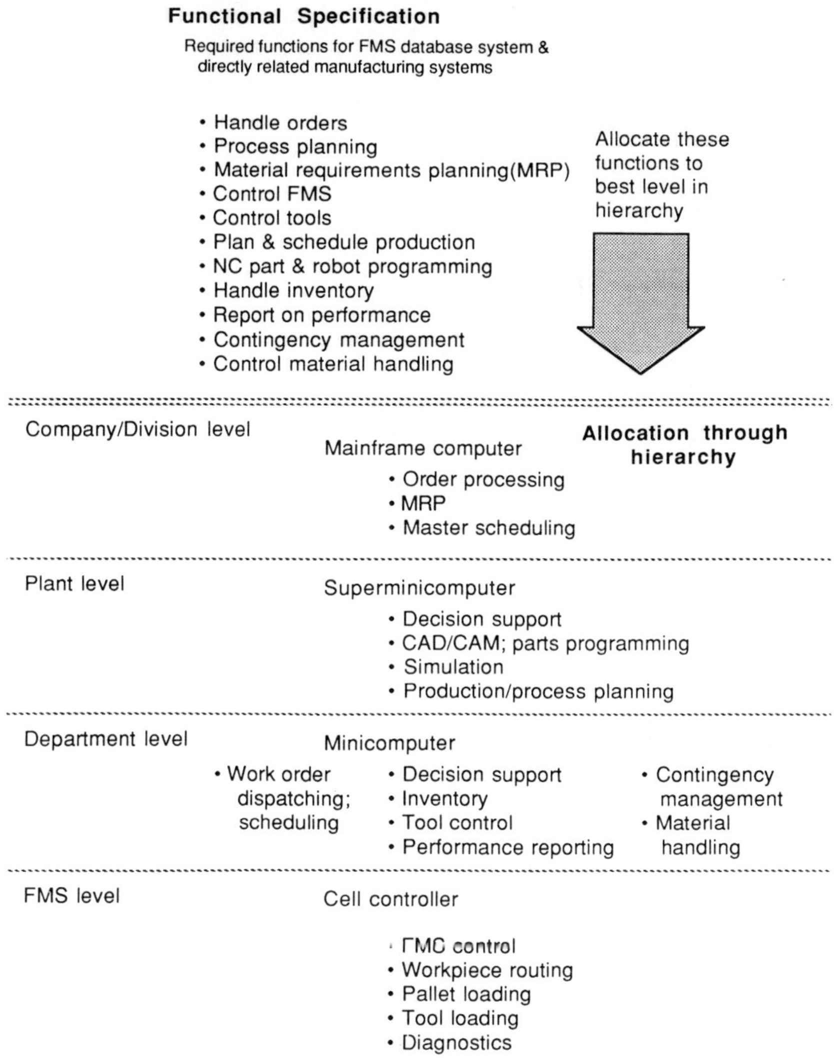

Figure 11. Functions developed in functional specification are allocated down through the system hierarchy.

controllers) to control everything, but such a system tends to suffer from many start-up difficulties and unreliability due to its extreme complexity.

Figure 12 shows a possible system design for the functional hierarchy of Figure 11, using two LANs. Note that the office/engineering network is separate from the factory network. This isolation prevents factory-type problems (e.g.,

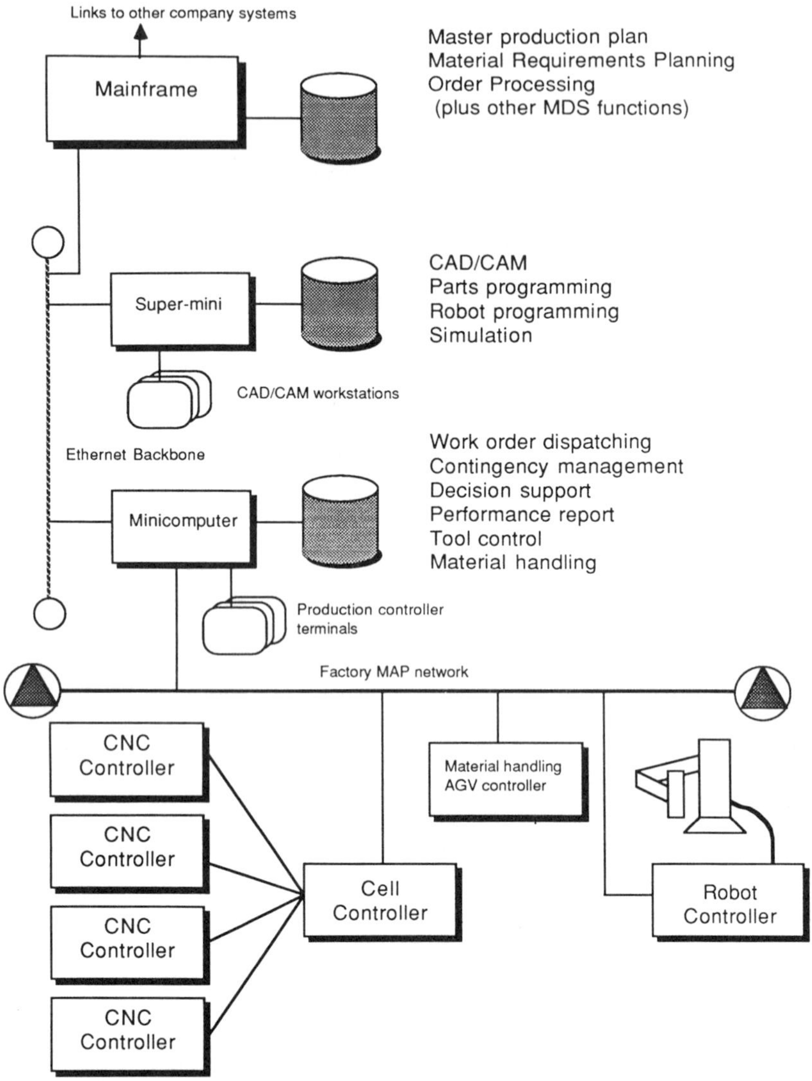

Figure 12. A system based on the hierarchy of Fig. 11.

LAN downtime due to welding next to a remodulator) from immediately affecting the entire network.

Design of a database layout and data processing operation for any type of system begins with understanding each function that deals with information. Determine

- *Input*. What information is required by this operation in order for it to perform its function? What information is stored at this function? What is the timing?
- *Processing*. How is information handled by this function? What information is originated at this function? Timing?
- *Output*. What information is supplied by this function to be used elsewhere? Timing?

As a first step, develop written documentation detailing the above functions. It is important to distinguish (*a*) what is actually being done now from (*b*) what *should* be in practice now, but isn't, and (*c*) how things will be done under a new, more computerized system.

As the required functionality starts to become clearer, it is then appropriate to identify interfaces between systems, content of databases at each level, and their frequency of update. Design of the logical structure of databases needs to be accomplished early in preliminary design, as program/procedure design depends upon it. Database structure, as mentioned earlier, is an overlay of the existing (or planned) business and manufacturing information structure. This is often dictated by physical realities (such as the relationship of individual tools to a machine) and data structure (e.g., the layout of scheduling data as customized to the manufacturing business in question).

When retrofitting FMS to an existing manufacturing operation, an evolutionary approach is superior to a revolutionary one. Trying to overhaul the entire process at once is likely to introduce so many problems simultaneously that it will be very difficult to determine conclusively where the bugs are. A phased approach, starting with the furthest downstream process (closest to the end) entails less risk and cost.

2. Relationship of the Flexible Manufacturing System Database to the Local Area Network

In a truly distributed database management system, there will be cases where a program running on computer A will have to access records in a database residing on computer B. Software mechanisms exist to provide for this, but the network link then becomes critical to completion of processing. If a distributed DBMS is planned for the FMS system, LAN capacity and reliability take on an even higher priority.

This is also the case when considering robot control and the LAN. If the LAN is used to transfer the robot program from the departmental database to a local robot controller, it serves only an intermittent function. If, however, the robot is being controlled directly over the LAN, LAN response time and capacity can make or break robot operation.

D. OPERATION

1. Procedures and Timing

An FMS database operation is an application where a data processing function must be tightly linked to factory floor timing requirements. Written procedures must be developed accordingly, and a schedule developed to ensure that needed data processing operations are completed when required.

When starting up a new FMS database system, begin testing as early as possible in the development phase. This will require the use of program stubs to simulate the actions of software which has not yet been written. This will also provide a test of the procedures themselves and provide more opportunity for production personnel to gain familiarity with the new systems. Gradual loading of databases and testing with subsets of the production information is preferable to performing a massive database load and attempting to begin full processing all at once.

The FMS databases may be stable (changing infrequently, such as the capacity of a machine) or very volatile (changing frequently, such as a production schedule). The more volatile a database is, the greater its need for maintenance. In any case, each database must have a designated owner who takes responsibility for ensuring its correctness. Updates to volatile databases may be handled automatically, by downloading or processing of transactions from another system, but a human must still take responsibility for database integrity.

When planning to upgrade from a largely manual record-keeping system to an automated database system, it is important to remember this: If good data integrity methods have been developed, the computerized system will always be more accurate than the older manual system. However, at first, people will always assume the new database is in error. It is vital to pay close attention to database integrity at this point: Investigate and explain every discrepancy. Once the system's credibility is established and maintained, it will become a valuable company asset. If the system gains a reputation for constantly being inaccurate, its credibility will suffer a long-term loss and the database system will fail to serve its purpose.

2. Synchronization of Distributed Databases: Contingency Planning

A distributed database system will always have some planned redundancy. This is needed to allow some independent processing to take place, eliminate a central computer as a single point of failure, and enhance performance. Redundancy frequently means, however, that differences will arise between databases which are supposed to be identical (at least, identical in certain fields).

Procedures for reconciling discrepancies between databases include a number of factors. First, a master database (and its human owner) must be defined, which is assumed to be the "most correct." (Database design and maintenance procedures must also reflect this assumption.) When designing databases, the performance/redundancy trade-off must include a consideration of discrepancy reconciliation procedures. These procedures will contain rules defining how discrepancies are to be handled and frequency/methods of reconciliation. Planning must also include methods for coordinating updates to multiple databases containing redundancy. Finally, "Murphy's Law" contingency planning is needed for dealing with the effects of database corruption, disk crashes, and even the complete loss of a computer.

VII. Questions to Be Considered When Planning the Flexible Manufacturing System Database

- What is the best allocation of intelligence down through a hierarchy of functional flow of computer control system? What databases will be at what level? Effectiveness of operation will be strongly influenced by the communications method.
- Will there be planned data redundancy, and if so, how will reconciliation be handled? Will there be one master database which will be considered the "most correct?"
- Will there be manual inputs? If so, will there be time delays (computer awaiting human input) due to relatively light manpower in a highly automated factory?
- Have database operations been included in the simulation of FMS during its design stages?
- How much data must be sent over communication networks to support system operation? Can network capacity handle it?
- If a LAN is acting as a real-time link between a database system and production equipment, will system response time be quick enough? (Failure here can mean equipment/product damage.)
- Contingencies: If the database becomes quite inaccurate due to problems, glitches, and so on, how will FMS operation be restored? Develop detailed error recovery procedures.
- The combination of database complexity, software complexity, and computer performance can combine to result in operation slow enough to disrupt FMS operation. Has a sensitivity analysis been performed as part of system

design and simulation to determine where fast computer response times are required to maintain FMS throughput?

VIII. Case Study Example

An example use of databases in FMS is a prototype system for building wire harnesses jointly developed by the Boeing Company and the U.S. Army Missile Command (Figure 13).

A wire harness design system runs on an IBM-compatible personal computer (PC) under the control of the AutoCad CAD program. This system consists of several parts libraries (containing available harness components) and a design program. The design program allows the engineer to specify the desired characteristics of parts to be included in the wire harness, and then search the library for parts which meet the requirements. Selected components are then placed in a graphic design file (the geometry portion of the design database). The design program forces the design engineer to consider producibility constraints during product design, as designs which exceed the FMS's capabilities cannot be developed using the design program (these can be overridden by using stock features of the CAD system, if desired, but it takes an overt effort). As the design file is constructed, a database is also built by the software which contains the characteristics and locations of each of the components. This database corresponds to the attribute portion of the design database.

When the wire harness design is completed, the engineer runs a data extraction program on the PC. At that time, information describing the harness is collated into a file and sent to the department VAX for processing.

In the VAX, the file is decomposed into several files, each of which is pertinent to a manufacturing step (i.e., wire cutting, stripping, terminating, layup, and plugging). These files are prepared to act as information for the controllers which will direct the various steps of the build process. These files reside on the VAX until the build schedule calls for this specific wire harness to be built.

The build files are then sent to the flexible manufacturing cell (FMC) controller. When the FMC manufacturing operation is started, the FMC controlled distributes a file to each step controller which the controller uses (as data) to perform its function in the harness build process. (Note that the controller software itself is unchanging; only the build data differ from harness to harness.)

This system is experimental but demonstrates the technical feasibility of complete integration from CAD to FMS. A key feature is the inclusion of high levels of producibility intelligence in the design program.

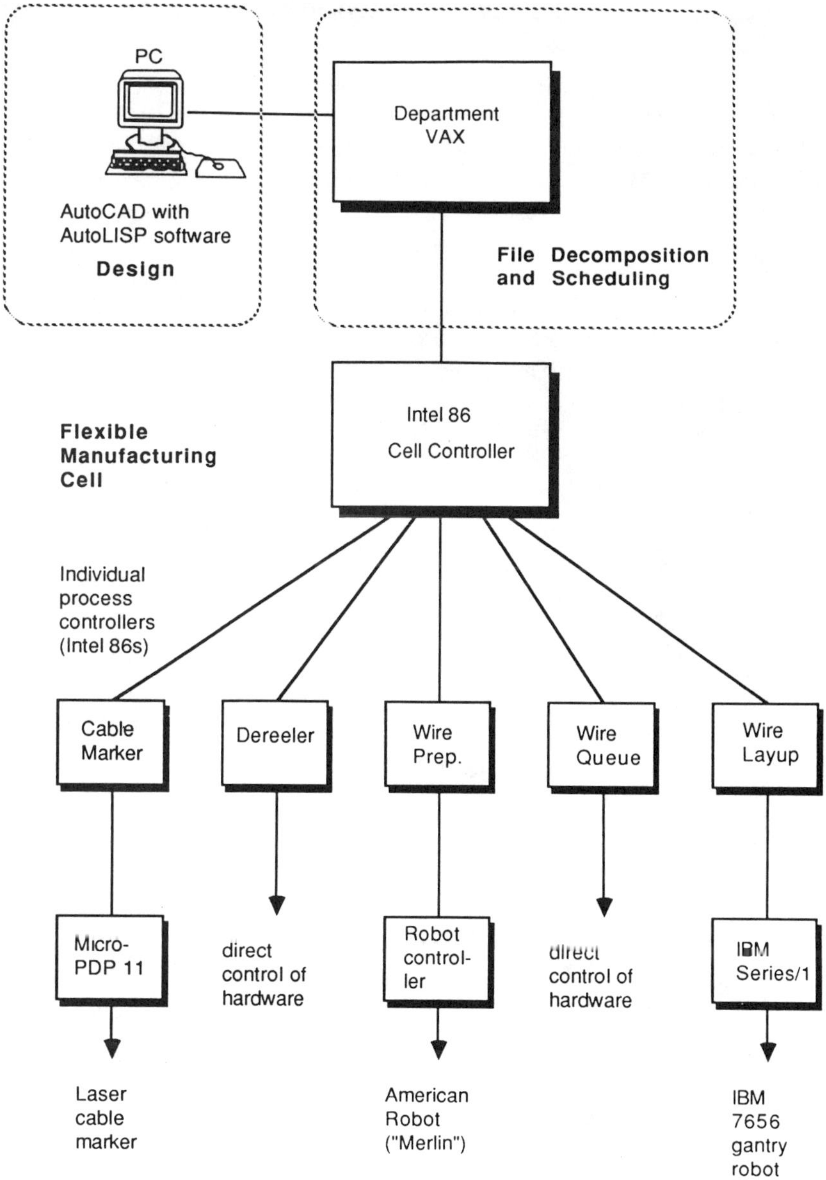

Figure 13. FMS database flow for wire harness assembly system.

IX. Conclusions

Databases and the software that interacts with them are a key element of a flexible manufacturing system. A systems approach must be followed to integrate FMS databases and systems with the rest of the manufacturing data system. Well-established system design methodologies are available to aid in this task. System design for FMS database systems includes more than just database layouts, however. Requirements for system operation imposed by realities of the manufacturing environment include considerations of speed and allocation of functionality through a manufacturing hierarchy which may include everything from mainframes to programmable controllers. Also, staffing and organizational considerations must be taken into account to ensure effective data preparation for the needs of FMS.

References

Engelke, W. D. (1987) "How to Integrate CAD/CAM Systems, Management and Technology," p. 30. Dekker, New York.

Greenwood, N. R. (1988a). "Implementing Flexible Manufacturing Systems," p. 158. Wiley, New York.

Greenwood, N. R. (1988b). "Implementing Flexible Manufacturing Systems," p. 174. Wiley, New York.

Further Readings

Association for Integrated Manufacturing Technology (1986). "A New Beginning," 23rd Annual Tech. Conf. Proc., May 4–7, 1986. AIM Tech, Beloit, Wisconsin.

Chan, T. S., and Pak, H. A. (1986). Heuristical job allocation in a flexible manufacturing system. *Int. J. Adv. Manuf. Technol.* **1**(2), 69–90.

Charharbargi, K., and Davies, B. L. (1986). The analysis of flexible manufacturing systems for reliability. *Int. J. Adv. Manuf. Technol.* **1**(4), 79–100.

Charles Stark Draper Laboratory, Inc. (1983). "Flexible Manufacturing System Handbook." U.S. Department of Commerce, National Technical Information Service, Warren, Michigan.

Choobineh, F., and Suri, R., eds. (1986). "Flexible Manufacturing Systems, Current Issues." Industrial Engineering and Management Press, Norcross, Georgia.

Chorafas, D. N. (1983). "DBMS for Distributed Computers and Networks." Petrocelli Books, Inc., New York.

Cowan, D., and Kochan, A. (1986). "Implementing CIM." Springer-Verlag, Berlin.

Crestin, J. P., and McWaters, J. F., eds. (1986). "Software for Discrete Manufacturing." North-Holland Publ., Amsterdam.

Engelke, W. D. (1987). "How to Integrate CAD/CAM Systems: Management and Technology." Dekker, New York.

Engelke, W. D., Minor, D. C., and Morgan, T. D. (1988). "Wire Harness CAD/CAM Interface Development," UAH Res. Rep. No. 691. University of Alabama, Huntsville.

Fabryky, W. J., and Blanchard, B. S. (1981). "Systems Engineering and Analysis." Prentice-Hall, Englewood Cliffs, New Jersey.

Flores, I. (1981). "Data Base Architectures." Van Nostrand-Reinhold, New York.

Ford, S. J. (1984). CIM's bridge from CADD to CAM. *IPAD II, Adv. Distrib. Data Base Manage. CAD/CAM, Proc. Nat. Symp.*, Denver, Colorado, *1984* pp. 28–29.

Greenwood, N. R. (1988). "Implementing Flexible Manufacturing Systems." Wiley, New York.

Groover, M. P., and Zimmers, E. W. (1984). "CAD/CAM." Prentice-Hall, Englewood Cliffs, New Jersey.

Hutchinson, G. K., and Chaturvedi, A. (1987). Information organization in flexible automation systems. *Comput. Ind.* **9.**

IBM Corporation (1983). "A Manager's Guide to CADAM System Implementation." IBM Corporation, White Plains, New York.

Knox, C. (1984). "Engineering Documentation for CAD/CAM Applications." Dekker, New York.

Kusiak, A., ed. (1986). "Flexible Manufacturing Systems: Methods and Studies." North-Holland Publ., Amsterdam.

Kusiak, A., ed. (1986). "Modelling and Design of Flexible Manufacturing Systems." Elsevier, Amsterdam.

Kusiak, A. (1987). Artificial intelligence and operations research in flexible manufacturing systems. *INFOR* **25**(1), 2–12.

Le Mastre, C., and El-Sawy, A. (1987). "Computer Integrated Manufacturing: A Systems Approach." UNIPUB, White Plains, New York.

Nazemetz, J. W., Hammer, W. E., and Sadowski, R. P., ed. "Computer Integrated Manufacturing Systems: Selected Readings." Industrial Engineering and Management Press, Norcross, Georgia.

Oden, H. W. (1987). Integrating manufacturing resources planning (MRP II) with flexible manufacturing systems (FMS). *Compt. Ind. Eng.* **13** (1–4), 107–111.

Preston, E. J., Crawford, G. W., and Coticchia, M. E. (1984). "CAD/CAM Systems, Justification, Implementation, Productivity Measurement." Dekker, New York.

Rathmill, K., ed. (1987) "Proceedings of the Fifth International Conference on Flexible Manufacturing Systems, 3–5 November 1986." Stratford-upon-Avon, U.K.

Stover, R. (1984). "An Analysis of CAD/CAM Applications." Prentice-Hall, Englewood Cliffs, New Jersey.

Tou, J. T., ed. (1985). "Computer-Based Automation." Plenum, New York.

Williams, D. J. (1988). "Manufacturing Systems: An Introduction to the Technologies." Wiley, New York.

CHAPTER 6

Group Technology in Flexible Manufacturing Systems

Andrew Kusiak
Department of Industrial Engineering
The University of Iowa
Iowa City, Iowa

I. Introduction

The basic idea of group technology (GT) is to decompose a manufacturing system into subsystems. Introduction of GT in manufacturing has the following advantages (Kusiak, 1990):

1. Reduced production lead time (20–88%)
2. Reduced work-in-process (up to 88%)

3. Reduced labor (15–25%)
4. Reduced tooling (20–30%)
5. Reduced rework and scrap materials (15–75%)
6. Reduced setup time (20–60%)
7. Reduced order time delivery (13–136%)
8. Improved human relations
9. Reduced paper work

One may question whether the cellular concept is applicable to flexible manufacturing systems. Some studies have shown that grouping machines into machine cells might limit the manufacturing system flexibility. However, industrial applications have proven that it is virtually impossible to implement a large-scale automated manufacturing system without using the cellular concept. There are at least four crucial factors which support this thesis:

• *Volume of information:* Volume of information in a large-scale automated manufacturing system is typically large and it is too expensive to effectively process information without the system's decomposition.

• *Material handling system:* In a typical automated manufacturing system, automated material handling carriers are used, for example automated guided vehicles (AGVs) or robots. Each of the two carriers tends a limited number of machines. This limitation for AGVs is imposed by the following:

1. Maximum number of trips that an AGV can make per time unit.
2. AGV guidance path, for example, a wire, which one should avoid to intersect by another guidance path within the AGV's working area. An AGV while approaching an intersection slows down and this in turn reduces its utilization.

• *Technological requirements:* Some machines have to be grouped together due to technological requirements, for example, a forging machine and a heat treatment station.

• *Management:* Although in most of the currently operating automated manufacturing systems, the degree of automation is higher than in the classical manufacturing systems, humans will be, for a long time, part of these new manufacturing systems. Due to the limited size of each machine cell, a cellular manufacturing system is easier to manage than a functionally organized system. It seems that this view has been given especially serious consideration in Europe and Japan.

The initial sections of this chapter focus on models and algorithms developed for GT. There are two basic methods used for solving the GT problem: classification and cluster analysis.

A. CLASSIFICATION METHOD

The classification method is used to group parts into part families based on their design features. There are two variations of the classification method: visual method and coding method.

1. Visual Method

The visual method is a semisystematic procedure where parts are grouped according to the similarity of the geometric shape, as shown in Figure 1, where 10 parts have been grouped into 4 part families. Grouping parts using the visual method is dependent on the personal preference. Therefore, this method is applicable in cases where the number of parts is rather limited.

2. Coding Method

In the coding method, parts can be classified on the basis of the following features:

1. Geometric shape and complexity
2. Dimensions
3. Type of material
4. Shape of raw material
5. Required accuracy of the finished part

Using a coding system, each part is assigned a numerical and/or alphabetical code. Each digit of this code represents a feature of a part. The currently

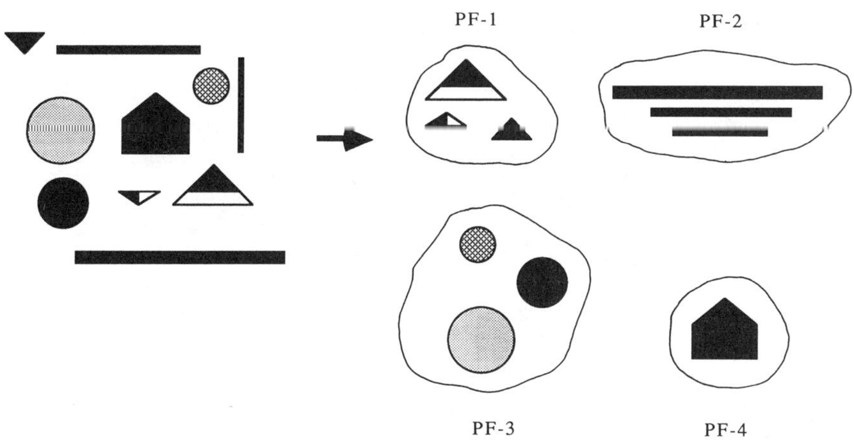

Figure 1. Parts grouped into families using a visual method.

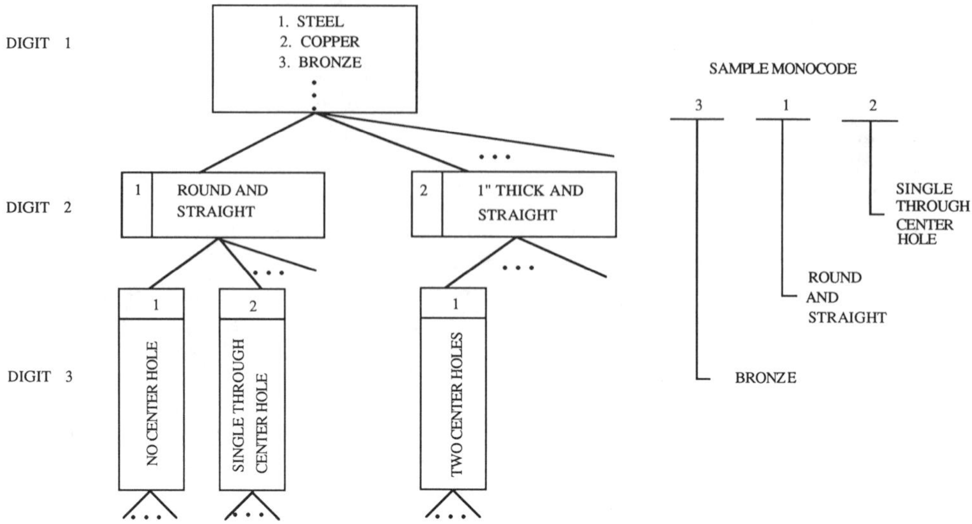

Figure 2. A monocode system and a monocode.

available coding systems differ with respect to the depth of coverage of the preceding five features. For example, a coding system may provide more information on the shape and dimension of a part, whereas another may emphasize accuracy of the part.

There are three basic types of coding systems: monocode, polycode, and hybrid. An example of a monocode system is illustrated in Figure 2. Features of each part are matched with the list of features corresponding to each node of the three and a part code is generated. Since the monocode system has a tree structure, a digit selected at a particular node depends on the digit selected at the preceding node. To fully understand the representation of a part monocode, all its digits are required. For a given part the length of its monocode is rather short compared to other coding systems (Ingram, 1982).

B. CLUSTER ANALYSIS METHOD

Cluster analysis is concerned with the grouping of objects into homogeneous clusters (groups) based on the object features. It has been applied in many areas such as: biology (Everitt, 1980), data recognition (McCormick *et al.*, 1972), medicine (Klastorin, 1982), pattern recognition (Tou and Gonzalez, 1974), production flow analysis (Burbidge, 1971; King, 1980), control engineering (Siljak and Sezer, 1984), automated systems (Kusiak, 1985; Kumar *et al.*, 1986), and expert systems (Cheng and Fu, 1985).

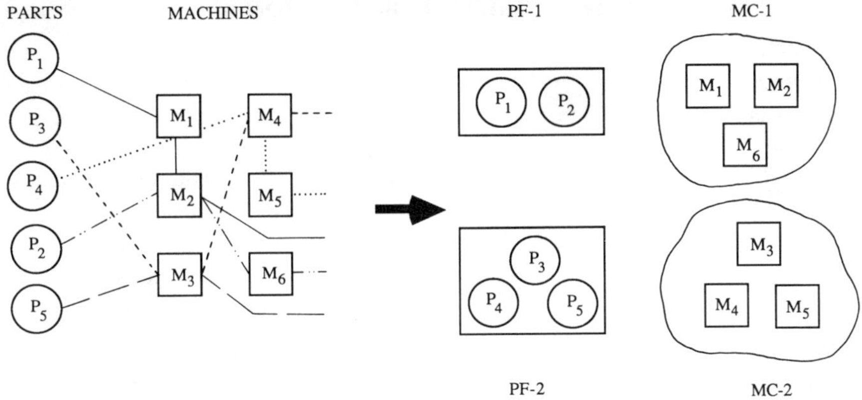

Figure 3. The physical machine layout.

The application of cluster analysis in GT is to group parts into part families (PF) and machines into machine cells (MC). The result of this grouping leads to physical machine layout or logical machine layout. The physical machine layout requires rearrangement of machines so that the shop floor is altered, as shown in Figure 3, whereas, using the logical machine layout, machines are grouped into logical machine cells and the position of machines is not altered (Figure 4). The logical grouping can be applied in a case when the production content is changing frequently so that the physical machine layout is not justified.

To model the GT problem, three formulations are used:

- Matrix formulation
- Mathematical programming formulation
- Graph formulation

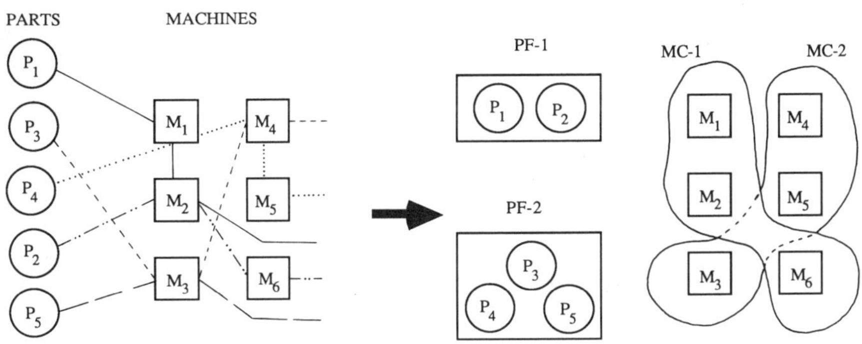

Figure 4. The logical machine layout.

II. Matrix Formulation

In the matrix formulation, a machine-part incidence matrix $[a_{ij}]$ is constructed. The machine-part incidence matrix $[a_{ij}]$ consists of 0,1 entries, where an entry 1 (0) indicates that machine i is used (not used) to process part j. Typically, when an initial machine-part incidence matrix $[a_{ij}]$ is constructed, clusters of machines and parts are not visible. Clustering algorithms allows one to transform an initial incidence matrix into a more structured (possibly block diagonal) form. To illustrate the clustering concept, consider the machine-part incidence Matrix (1).

$$
\begin{array}{c}
\text{part number} \\
\begin{array}{ccccc} 1 & 2 & 3 & 4 & 5 \end{array}
\end{array}
$$

$$
[a_{ij}] = \begin{array}{c} 1 \\ 2 \\ 3 \\ 4 \end{array}\begin{bmatrix} & 1 & & 1 & 1 \\ 1 & & 1 & & \\ & 1 & & 1 & \\ 1 & & 1 & & \end{bmatrix} \begin{array}{l} \text{machine} \\ \text{number} \end{array} \qquad (1)
$$

Rearranging rows and columns in Matrix (1) results in Matrix (2)

$$
\begin{array}{ccc}
 & & \begin{array}{ccccc} \multicolumn{2}{c}{\text{PF-1}} & \multicolumn{3}{c}{\text{PF-2}} \\ 1 & 3 & 2 & 4 & 5 \end{array} \\
\begin{array}{c} \text{MC-1} \\ \\ \text{MC-2} \\ \\ \end{array} &
\begin{array}{c} 2 \\ 4 \\ 1 \\ 3 \end{array} &
\left[\begin{array}{cc|ccc} 1 & 1 & & & \\ 1 & 1 & & & \\ \hline & & 1 & 1 & 1 \\ & & 1 & 1 & \end{array}\right]
\end{array} \qquad (2)
$$

Two machine cells (clusters) MC-1 = {2, 4}, MC-2 = {1, 3}, and two corresponding part families PF-1 {1, 3}, PF-2 = {2, 4, 5} are visible in Matrix (2).

Clustering of a binary incidence matrix may result in the following two categories of clusters: (1) mutually separable and (2) partially separable. The two categories of clusters are presented in Matrices (2) and (3).

$$
\begin{array}{ccc}
 & & \begin{array}{ccccc} 1 & 2 & 3 & 4 & 5 \end{array} \\
\begin{array}{c} \\ \text{MC-1} \\ \\ \text{MC-2} \end{array} &
\begin{array}{c} 1 \\ 2 \\ 3 \\ 4 \end{array} &
\left[\begin{array}{cc|cc|c} 1 & 1 & & & 1 \\ 1 & 1 & & & \\ \hline & & 1 & 1 & 1 \\ & & 1 & 1 & \end{array}\right]
\end{array} \qquad (3)
$$

Matrix (3) cannot be separated into two disjoint clusters because of part 5, which is to be machined in two cells MC-1 and MC-2. Removing part 5 from Matrix (3) results in the decomposition of Matrix (3) into two separable machine cells MC-1 = {1, 2}, MC-2 = {3, 4} and two part families PF-1 = {1, 2} and

PF-2 = {3, 4}. The two clusters are called partially separable clusters. To deal with the bottleneck part 5, one of the following three actions can be taken:

1. It can be machined in one machine cell and transferred to the other machine cell by a material handing carrier.
2. It can be machined in a functional facility.
3. It can be subcontracted.

To solve the matrix formulation of the group technology problem the following approaches have been developed:

- Similarity coefficient methods
- Sorting based algorithms
- Bond energy algorithm
- Cost-based method
- Cluster identification algorithm
- Extended cluster identification algorithm

A. SIMILARITY COEFFICIENT METHODS

In this section, three similarity coefficient methods are discussed. The single linkage cluster analysis (SLCA) presented by McAuley (1972) is based on the similarity coefficient s_{ij} measure between two machines i and j and is computed as follows:

$$s_{ij} = \frac{\sum_{k=1}^{n} \delta_1 (a_{ik}, a_{jk})}{\sum_{k=1}^{n} \delta_2 (a_{ik}, a_{jk})}$$

where

$$\delta_1 (a_{ik}, a_{jk}) = \begin{cases} 1 \text{ if } a_{ik} = a_{jk} = 1 \\ 0 \text{ otherwise} \end{cases}$$

$$\delta_2 (a_{ik}, a_{jk}) = \begin{cases} 0 \text{ if } a_{ik} = a_{jk} = 1 \\ 1 \text{ otherwise} \end{cases}$$

To solve the GT problem using the SLCA approach, similarity coefficients for all possible pairs of machines are computed. Machine cells are generated based on a threshold value of the similarity coefficient.

To illustrate the SLCA, consider Matrix (3). The similarity coefficients s_{ij} are computed below and depicted in Figure 5.

$$s_{12} = s_{34} = \frac{2}{2 + 1} = \frac{2}{3} = 0.75$$

$$s_{13} = \frac{1}{4 + 1} = \frac{1}{5} = 0.25$$

$$s_{14} = s_{23} = s_{24} = \frac{0}{5} = 0$$

Assuming the threshold value of the similarity coefficient $s_{ij} = 60\%$, from Figure 5 the following machine cells are obtained: MC-1 = {1, 2} and MC-2 = {3, 4}.

One of the disadvantages of the SLCA is that it fails to recognize the chaining problem resulted from the duplication of bottleneck machines. A bottleneck machine is one that does not allow one to decompose a machine-part incidence matrix into submatrices. For example, machine 3 in Matrix (4) does not permit the decomposition of that matrix into two machine cells and two part families.

$$\text{machine}\;\begin{matrix}1\\2\\3\\4\\5\end{matrix}\overset{\text{part}}{\overset{\begin{matrix}1&2&3&4&5&6\end{matrix}}{\begin{bmatrix}1&1&&&&\\1&1&&&&\\1&1&1&&1&1\\&&1&1&1&1\\&&1&&1&1\end{bmatrix}}} \tag{4}$$

A way to decompose Matrix (4) into two mutually separable submatrices is to purchase an additional copy of machine 3. The latter allows one to transform Matrix (4) into Matrix (5).

$$\begin{matrix}\text{MC-1}\begin{cases}1\\2\\3\end{cases}\\ \text{MC-2}\begin{cases}3\\4\\5\end{cases}\end{matrix}\overset{\overbrace{\begin{matrix}1&2\end{matrix}}^{\text{PF-1}}\;\overbrace{\begin{matrix}3&4&5&6\end{matrix}}^{\text{PF-2}}}{\left[\begin{array}{cc|cccc}1&1&&&&\\1&1&&&&\\1&1&&&&\\\hline&&1&&1&1\\&&1&1&1&1\\&&1&&1&1\end{array}\right]} \tag{5}$$

Two machine cells MC-1 = { 1, 2, 3 }, MC-2 = { 3, 4, 5 } and two corresponding part families PF-1 = { 1, 2 }, PF-2 = { 3, 4, 5, 6 } are visible in Matrix (5). Analogously to the definition of a bottleneck machine, a bottleneck part can be defined.

Use of the average linkage clustering (ALC) algorithm to overcome the chaining problem was studied by Seifoddini and Wolfe (1986). They defined the similarity coefficient between any two clusters as an average of the similarity coefficient between all elements of the two clusters. To solve the group technology problem, Seifoddini and Wolfe represented the machine-part incidence

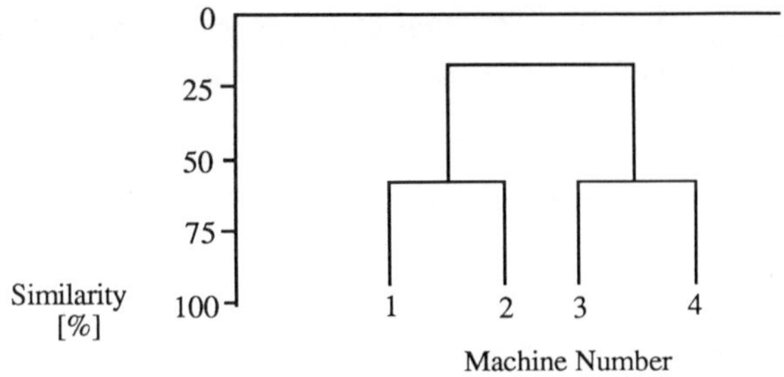

Figure 5. A tree of similarity coefficients.

matrix using a binary machine code. The total number of intercellular movements (ICM) between two machine cells MC-i and MC-j is computed as follows:

$$ICM_{ij} = \sum_{k=1}^{n} \delta_3 (v_{ik}, v_{jk})$$

where

$$v_{ik} = \begin{cases} 1 \text{ if } \sum_{i \in MC\text{-}i} a_{ik} \neq 0 \\ 0 \text{ otherwise} \end{cases}$$

$$\delta_3 (v_{ik}, v_{jk}) = \begin{cases} 1 \text{ if } v_{ik} = v_{jk} = 1 \\ 0 \text{ otherwise} \end{cases}$$

Note that $\delta_3 (v_{ik}, v_{jk}) = 1$ indicates that part k requires processing in both machine cells MC-i and MC-j, and thus part k is a bottleneck part. This problem can be solved either by removing part k from the two machine cells or adding an identical machine, say q, to each machine cell. The average linking clustering algorithm is illustrated below.

Consider the example illustrated in Figure 3 where two clusters were obtained. The intercellular movement is computed as follows:

$$[v_{1k}] = [1, 1, 0, 0, 1]$$

$$[v_{2k}] = [0, 0, 1, 1, 1]$$

$$ICM_{12} = \sum_{k=1}^{5} \delta_3 (v_{1k}, v_{2k}) = 0 + 0 + 0 + 0 + 1 = 1$$

The only part which requires processing in both machine cells MC-1 and MC-2 is part 5 with corresponding $\delta_3 (v_{15}, v_{25}) = 1$.

Another interesting similarity coefficient method was studied by De Witte

(1980). He designed a clustering algorithm advocating the concept of some machines to be included in more than one machine cell. He divided all the available machines into (1) primary machines, (2) secondary machines, and (3) tertiary machines. To analyze the relationship between these machines three different similarity coefficients were used:

1. Absolute similarity coefficient sa_{ij}
2. Mutual similarity coefficient sm_{ij}
3. Single similarity coefficient ss_{ij}

De Witte concluded that the best way of clustering is to begin with coefficient sa_{ij} and sm_{ij} and then use ss_{ij} to allocate the remaining unassigned machines.

B. SORTING-BASED ALGORITHMS

The method of clustering based on sorting rows and columns of the machine-part incidence matrix has been studied by many authors. King (1980) developed the rank order clustering (ROC) algorithm.

ROC Algorithm

Step 1. For each row of the machine-part incidence matrix, assign a binary weight and calculate a decimal equivalent (a weight).

Step 2. Sort rows of the binary matrix in decreasing order of the corresponding decimal weights.

Step 3. Repeat the above two steps for each column.

Step 4. Repeat the above steps until the position of each element in each row and column does not change. A weight for each row i and column j is calculated as follows:

$$\text{row } i \quad : \sum_{k=1}^{n} a_{ik} 2^{n-k}$$

$$\text{column } j : \sum_{k=1}^{m} a_{kj} 2^{n-k}$$

In the final matrix generated by the ROC algorithm, clusters are identified visually. The ROC algorithm for Matrix (1) is illustrated in Example 1.

Example 1

Step 1. Assign binary weights to each row and calculate decimal equivalents.

	Part 1	Part 2	Part 3	Part 4	Part 5	Decimal equivalent	Machine number
binary weight	2^4	2^3	2^2	2^1	2^0		
		1		1	1	11	1
	1		1			20	2
		1		1		10	3
	1		1			20	4

Step 2. Sorting the decimal weights in decreasing order results in the following matrix:

Machine number	Part 1	Part 2	Part 3	Part 4	Part 5
2	1		1		
4	1		1		
1		1		1	1
3		1		1	

Step 3: Repeating the above steps for each column produces the following matrix:

Machine number	Part 1	Part 3	Part 2	Part 4	Part 5
2	1	1			
4	1	1			
1			1	1	1
3			1	1	

In the above matrix two separable clusters are visible. The matrix obtained is identical to Matrix (2).

The ROC algorithm was further extended in King and Nakornchai (1982) and Chandrasekharan and Rajagopalan (1986). In the latter paper, Chandrasekharan and Rajagopalan improved the ROC algorithm incorporating "block and slice" method and the hierarchical clustering method.

Another interesting sorting-based algorithm, the direct cluster algorithm (DCA), was developed by Chan and Milner (1982). The DCA consists of the following steps:

Step 1. Determine the total number of 1s in each row and column in the machine-part incidence matrix.

Step 2. Sort each row in increasing order corresponding to the total number of 1s.

Step 3. Sort each column in decreasing order corresponding to the total number of 1s.

Step 4. Repeat the above steps until the position of each element in each row and column does not change.

Based on the final outcome of the above steps, some rows and columns in the machine-part matrix are to be rearranged until satisfactory clusters are obtained.

C. BOND-ENERGY ALGORITHM

McCormick *et al.* (1972) developed an interchange clustering algorithm called the bond-energy algorithm (BEA). The BEA seeks to form a block diagonal form by maximizing the measure of effectiveness which is defined as follows:

$$\mathrm{ME} = 1/2 \sum_{i=1}^{m} \sum_{j=1}^{n} a_{ij} \left[a_{i,j-1} + a_{i,j+1} + a_{i-1,j} + a_{i+1,j}\right]$$

The ME is illustrated below.

(A)

$$\begin{array}{cc} & \text{part number} \\ & \begin{array}{cccc} 1 & 2 & 3 & 4 \end{array} \\ \text{machine number} \begin{array}{c} 1 \\ 2 \\ 3 \\ 4 \end{array} & \begin{bmatrix} 1 & & 1 & \\ & 1 & & 1 \\ 1 & & 1 & \\ & 1 & & 1 \end{bmatrix} \end{array}$$

ME = 0

(B)

$$\begin{array}{cc} & \text{part number} \\ & \begin{array}{cccc} 1 & 2 & 3 & 4 \end{array} \\ \text{machine number} \begin{array}{c} 1 \\ 2 \\ 3 \\ 4 \end{array} & \begin{bmatrix} 1 & 1 & & \\ & & 1 & \\ 1 & 1 & & 1 \\ & & 1 & 1 \end{bmatrix} \end{array}$$

ME = 4

BEA Algorithm

Step 1. Set $j = 1$. Select one of the columns arbitrarily.

Step 2. Place each of the remaining $n - j$ columns, one at a time, in each $j + 1$ position, and compute each column's contribution to the ME. Place the column that gives the largest incremental contribution to the ME in its best location. Increase j by 1 and repeat the above steps until $j = n$.

Step 3. When all the columns have been placed, repeat the procedure for the rows.

The bond-energy algorithm applied to Matrix (1) is illustrated in Example 2.

Example 2

Step 1. Set $j = 1$. Select column 2.

Step 2. Placing each of the remaining columns in each of the $j + 1$ positions, its contribution to the ME value is computed below:

Position	j = 1	j + 1	ME value
	2	1	0
Column	2	3	0
number	2	4	2
	2	5	1

Column 4 is placed in the $j + 1$ position. Repeating the same procedure for the remaining columns leads to the following column order: {2, 4, 5, 1, 3}.

Step 3. Repeating the above steps for rows results in the following matrix:

machine number \ part number	2	4	5	1	3
1	1	1	1		
3	1	1			
2				1	1
4				1	1

The value of ME for the above matrix is 9.

Slagle *et al.* (1975) developed a clustering algorithm based on the BEA and the shortest spanning path (SPS) algorithm. Their concept was then extended in Bhat and Haupt (1976). They developed an algorithm where the matching between any two rows (columns) of the incidence matrix is calculated as follows:

$$m_{ij} = \sum_{k=1}^{n} |a_{ik} - a_{jk}|$$

The Bhat and Haupt (1976) algorithm is similar to the McCormick *et al.* (1972) algorithm. The difference between the two algorithms is that the Bhat and Haupt algorithm permutes rows and columns of matrix $A \cdot A^T$ as opposed to permuting rows and calculating matchings in matrix A in the McCormick *et al* algorithm. One can note that an entry m_{ij} of matrix $[m_{ij}]$ represents a distance between rows i and j in matrix A.

D. COST-BASED METHOD

Askin and Subramanian (1987) developed a clustering algorithm which considers the following manufacturing costs:

1. Fixed and variable machining cost
2. Setup cost
3. Production cycle inventory cost

4. Work-in-process inventory cost
5. Material handling cost

The algorithm consists of three stages. In the first stage, parts are classified using a coding system. In stage two, an attempt is made to develop a feasible grouping of parts based on the manufacturing costs. In stage three, the actual layout of machine cells is analyzed.

E. CLUSTER IDENTIFICATION ALGORITHM

Kusiak and Chow (1987a) applied the concept presented in Iri (1968) to develop the cluster identification (CI) algorithm. The cluster identification algorithm allows one to check the existence of mutually separable clusters in a binary machine-part incidence matrix provided that they exist.

CI Algorithm

Step 0. Set iteration number $k = 1$.

Step 1. Select any row i of incidence matrix $A^{(k)}$ and draw horizontal line h_i through it ($A^{(k)}$ reads: matrix A at iteration k).

Step 2. For each entry "1" crossed by the horizontal line h_i draw a vertical line v_j.

Step 3. For each entry "1" crossed once by a vertical line v_j draw a horizontal line h_k.

Step 4. Repeat steps 2 and 3 until there are no more crossed once entries "1" in $A^{(k)}$. All crossed twice entries "1" in $A^{(k)}$ form machine cell MC-k and part family PF-k.

Step 5. Transform the incidence matrix $A^{(k)}$ into $A^{(k+1)}$ by removing rows and columns corresponding to all the horizontal and vertical lines drawn in steps 1 to 4.

Step 6. If matrix $A^{(k+1)} = \mathbf{0}$ (where $\mathbf{0}$ denotes a matrix with all elements equal to zero), stop; otherwise set $k = k + 1$ and go to step 1.

Note that the cluster identification algorithm scans each element of matrix A two times. Since there are mn elements in matrix A, its time computational complexity is $O(mn)$, that is, the number of elementary computer operations is of the order of mn.

Application of the cluster identification algorithm for solving the group technology problem is illustrated in Example 3.

Example 3

Consider the following machine-part incidence Matrix (6).

$$
\begin{array}{c} \\ \\ A = \end{array}
\begin{array}{c|cccccccc|}
 & \multicolumn{8}{c}{\text{part number}} \\
 & 1 & 2 & 3 & 4 & 5 & 6 & 7 & 8 \\
\hline
1 & & 1 & 1 & & 1 & & & \\
2 & 1 & & & & & 1 & & \\
3 & & & & 1 & & & 1 & \\
4 & 1 & & & & & 1 & & \\
5 & & & 1 & & 1 & & & 1 \\
6 & & & & 1 & & & & \\
7 & & 1 & 1 & & 1 & & & 1 \\
\hline
\end{array}
\begin{array}{l} \text{machine} \\ \text{number} \end{array}
\qquad (6)
$$

Step 0. Set iteration number $k = 1$.

Step 2. Row 1 of Matrix (6) is selected and horizontal line h_1 is drawn. The result of step 1 and step 2 is presented in Matrix (7).

Step 2. Three vertical lines v_2, v_3, and v_5 are drawn.

$$
A^{(1)} =
\begin{array}{c|cccccccc|l}
 & 1 & 2 & 3 & 4 & 5 & 6 & 7 & 8 & \\
\hline
1 & - & 1 & 1 & - & 1 & - & - & - & -h_1 \\
2 & 1 & \vdots & \vdots & & \vdots & 1 & & & \\
3 & & \vdots & \vdots & 1 & \vdots & & 1 & & \\
4 & 1 & \vdots & \vdots & & \vdots & 1 & & & \\
5 & & \vdots & 1 & & 1 & & & 1 & \\
6 & & \vdots & \vdots & 1 & \vdots & & & & \\
7 & & 1 & 1 & & 1 & & & 1 & \\
\hline
 & & v_2 & v_3 & & v_5 & & & &
\end{array}
\qquad (7)
$$

As a result of drawing these three vertical lines, five new crossed-once entries "1" are created in Matrix (7), that is, entries (5, 3), (5, 5), and entries (7, 2), (7, 3), and (7, 5).

Step 3. Two horizontal lines h_5 and h_7 are drawn through the crossed-once entries of Matrix (7), as shown in Matrix (8).

$$
A^{(1)} =
\begin{array}{c|cccccccc|l}
 & 1 & 2 & 3 & 4 & 5 & 6 & 7 & 8 & \\
\hline
1 & - & 1 & 1 & - & 1 & - & - & - & -h_1 \\
2 & 1 & \vdots & \vdots & & \vdots & 1 & & & \\
3 & & \vdots & \vdots & 1 & \vdots & & 1 & & \\
4 & 1 & \vdots & \vdots & & \vdots & 1 & & & \\
5 & - & + & 1 & - & 1 & - & - & 1 & -h_5 \\
6 & & \vdots & \vdots & 1 & \vdots & & & & \\
7 & - & 1 & 1 & - & 1 & - & - & 1 & -h_7 \\
\hline
 & & v_2 & v_3 & & v_5 & & & &
\end{array}
\qquad (8)
$$

Step 4. Since the entries (5, 8) and (7, 8) of Matrix (8) are crossed once, the vertical line v_8 is drawn as shown in Matrix (9).

$$
A^{(1)} = \begin{array}{c} \\ 1 \\ 2 \\ 3 \\ 4 \\ 5 \\ 6 \\ 7 \\ \\ \end{array}
\begin{array}{cccccccc}
1 & 2 & 3 & 4 & 5 & 6 & 7 & 8 \\
 & 1 & 1 & & 1 & & & \\
1 & & & & & 1 & & \\
 & & & 1 & & & 1 & \\
1 & & & & & 1 & & \\
 & & 1 & & 1 & & & 1 \\
 & & & 1 & & & & \\
 & 1 & 1 & & 1 & & & 1 \\
 & v_2 & v_3 & & v_5 & & & v_8
\end{array}
\begin{array}{c} \\ h_1 \\ \\ \\ \\ h_5 \\ \\ h_7 \\ \\ \end{array} \tag{9}
$$

Since there are no more crossed-once entries "1", all the crossed-twice entries "1" of Matrix (9) form machine cell MC-1 = {1, 5, 7} and part family PF-1 = {2, 3, 5, 8}.

Step 5. Matrix (9) is transformed into Matrix (10).

$$
A^{(2)} = \begin{array}{c} \\ 2 \\ 3 \\ 4 \\ 6 \end{array}
\begin{array}{cccc}
1 & 4 & 6 & 7 \\
1 & & 1 & \\
 & 1 & & 1 \\
1 & & 1 & \\
 & 1 & &
\end{array} \tag{10}
$$

In the second iteration ($k = 2$), steps 1 to 4 are performed on Matrix (10). This iteration results in incidence Matrix (11) and machine cell MC-2 = {2, 4} and part family PF-2 = {1, 6}.

$$
A^{(2)} = \begin{array}{c} \\ 2 \\ 3 \\ 4 \\ 6 \\ \\ \end{array}
\begin{array}{cccc}
1 & 4 & 6 & 7 \\
1 & & 1 & \\
 & 1 & & 1 \\
1 & & 1 & \\
 & 1 & & \\
v_1 & & v_6 &
\end{array}
\begin{array}{c} \\ h_2 \\ \\ h_4 \\ \\ \\ \end{array} \tag{11}
$$

In the third iteration ($k = 3$) Matrix (12) is generated.

$$
A^{(3)} = \begin{array}{c} \\ 3 \\ 6 \\ \\ \end{array}
\begin{array}{cc}
4 & 7 \\
1 & 1 \\
1 & \\
v_4 & v_7
\end{array}
\begin{array}{c} \\ h_3 \\ h_6 \\ \\ \end{array} \tag{12}
$$

From this matrix, MC-3 = {3, 6} and PF-3 = {4, 7} are obtained. The final clustering result is illustrated in Matrix (13). The layout corresponding to Matrix (13) is illustrated in Figure 6.

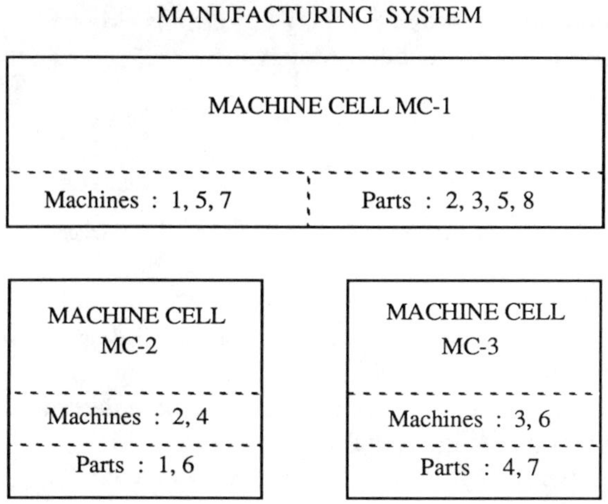

Figure 6. A layout of the manufacturing system from Example 2.

$$\begin{array}{ll} & \begin{array}{cccc|cc|cc} \multicolumn{4}{c}{\text{PF-1}} & \multicolumn{2}{c}{\text{PF-2}} & \multicolumn{2}{c}{\text{PF-3}} \\ 2 & 3 & 5 & 8 & 1 & 6 & 4 & 7 \end{array} \\ \begin{array}{l} \text{MC-1} \left\{ \begin{array}{c} 1 \\ 5 \\ 7 \end{array} \right. \\ \text{MC-2} \left\{ \begin{array}{c} 2 \\ 4 \end{array} \right. \\ \text{MC-3} \left\{ \begin{array}{c} 3 \\ 6 \end{array} \right. \end{array} & \left[\begin{array}{cccccccc} 1 & 1 & 1 & & & & & \\ & 1 & 1 & 1 & & & & \\ 1 & 1 & 1 & 1 & & & & \\ & & & & 1 & 1 & & \\ & & & & 1 & 1 & & \\ & & & & & & 1 & 1 \\ & & & & & & 1 & \end{array} \right] \end{array} \tag{13}$$

The computational experience showed that the cluster identification algorithm is very efficient. To show the efficiency of this algorithm, consider the clustering of a matrix $mxn = 60 \times 80$. It takes only 0.06 s to identify clusters by the developed algorithm. To cluster the same matrix by Bhat and Haupt (1976) algorithm, more than 300 s are required.

F. EXTENDED CLUSTER IDENTIFICATION ALGORITHM

Consider the following machine-part incidence Matrix (14).

$$\begin{array}{cc} & \text{part number} \\ & \begin{array}{ccccc} 1 & 2 & 3 & 4 & 5 \end{array} \\ \begin{array}{c} 1 \\ 2 \\ 3 \\ 4 \end{array} & \left[\begin{array}{ccccc} 1 & & 1 & 1 & \\ & 1 & 1 & & 1 \\ 1 & & 1 & & \\ & 1 & & & 1 \end{array} \right] \begin{array}{l} \\ \text{machine} \\ \text{number} \\ \\ \end{array} \end{array} \tag{14}$$

Using any of the existing clustering algorithms, Matrix (14) can be transformed into a more structured Matrix (15).

part number

	machine number	5	2	3	1	4
MC-1 {	2	1	1	1		
	4	1	1			
MC-2 {	1			1	1	1
	3			1	1	

(15)

Matrix (15) cannot be separated into two disjoint submatrices because of part 3, which is to be machined in two machine cells MC-1 and MC-2. Removing part 3 from Matrix (15) results in the decomposition of Matrix (15) into two mutually separable machine cells MC-1 = {2, 4}, MC-2 = {1, 3} and two part families PF-1 = {5, 2} and PF-2 = {1, 4}.

The extended cluster identification algorithm is developed to deal with overlapping parts. It takes advantage of both the expertise of a production manager regarding possible grouping of machines and parts and also the efficiency of the cluster identification algorithm.

Extended CI Algorithm

Step 0. Set iteration number $k = 1$.

Step 1. Select those machines (rows of matrix $A^{(k)}$) which, based on the user's expertise, are potential candidates for inclusion in machine cell MC-k. Draw horizontal lines h_i through each row of matrix $A^{(k)}$ corresponding to these machines. In the absence of the user's expertise any machine can be selected.

Step 2. For each column in $A^{(k)}$ corresponding to the entry "1" crossed by any of the horizontal lines h_i draw a vertical line v_j.

Step 3. For each row in $A^{(k)}$ corresponding to the entry "1" crossed by the vertical line v_j, drawn in step 2, draw a horizontal line h_i. Based on the machines corresponding to all the horizontal lines drawn in step 1 and step 3, a temporary machine cell MC-k is formed. If the production manager's expertise indicates that some of the machines cannot be included in the temporary machine cell MC-k', erase the corresponding horizontal lines in matrix $A^{(k)}$. Removal of these horizontal lines results in machine cell MC-k. Delete from matrix $A^{(k)}$ parts (columns) which are to be manufactured on at least one of the machines already included in MC-k. Place these parts on the list of parts to be manufactured in a functional machining facility. Draw a vertical line v_j through each crossed once entry "1" in $A^{(k)}$ which does not involve any other machines than those included in MC-k.

Step 4. For all the crossed twice entries "1" in $A^{(k)}$, form a machine cell MC-k and a part family PF-k.

Step 5. Transform the incidence matrix $A^{(k)}$ into $A^{(k+1)}$ by removing all the rows and columns corresponding to the rows and columns included in MC-k and PF-k, respectively.

Step 6. If matrix $A^{(k+1)} = \mathbf{0}$ (where $\mathbf{0}$ denotes a matrix with all elements equal to zero), stop; otherwise set $k = k + 1$ and go to step 1.

As can be seen, computational complexity of the extended cluster identification algorithm is slightly higher than the cluster identification algorithm, mainly due to step 3. In the case when the clustered matrix has the block diagonal structure, computational time complexity of the extended cluster identification algorithm reduces to $O(mn)$, that is, equal to the complexity of the cluster identification algorithm.

The extended cluster identification algorithm is illustrated in Example 4.

Example 4

Given the machine-part incidence, Matrix (16) determines mutually separable machine cells and part families.

$$A = \begin{array}{c} \\ 1\\2\\3\\4\\5\\6\\7 \end{array} \begin{array}{c} \text{part number} \\ \begin{array}{ccccccccccc} 1 & 2 & 3 & 4 & 5 & 6 & 7 & 8 & 9 & 10 & 11 \end{array} \\ \left[\begin{array}{ccccccccccc} & 1 & 1 & & & & 1 & & & & \\ 1 & & & & 1 & & & & & & 1 \\ & & & & & & & & & 1 & \\ 1 & & 1 & & & 1 & & & & & \\ & & & & 1 & & & 1 & & & \\ 1 & & & 1 & & & & 1 & 1 & 1 & \\ & & 1 & 1 & & 1 & 1 & & 1 & & \end{array}\right] \end{array} \begin{array}{c} \text{machine} \\ \text{number} \end{array} \qquad (16)$$

The user's expertise indicates that machines 1 and 4 should be included in the machine cell MC-1 and machine 3 and 5 should be included in the machine cell MC-2. Some other expertise required is indicated later in the example.

Step 0. Set iteration number $k = 1$.

Step 1. Since the user's expertise indicates that machines 1 and 4 should be included in machine cell MC-1, two horizontal lines h_1 and h_4 are drawn, as shown in Matrix (17).

$$A^{(1)} = \begin{array}{c} \\ 1 \\ 2 \\ 3 \\ 4 \\ 5 \\ 6 \\ 7 \end{array} \begin{array}{ccccccccccc} 1 & 2 & 3 & 4 & 5 & 6 & 7 & 8 & 9 & 10 & 11 \\ \hline & 1 & 1 & & & & 1 & & & & \\ 1 & & & & 1 & & & & & & 1 \\ & & & & & & & & & 1 & \\ \hline 1 & & 1 & & & 1 & & & & & \\ & & & & 1 & & & 1 & & & \\ 1 & & & 1 & & & & 1 & 1 & 1 & \\ & & 1 & 1 & & 1 & 1 & & 1 & & \end{array} \begin{array}{l} \\ \text{-- } h_1 \\ \\ \\ \text{-- } h_4 \\ \\ \\ \\ \end{array} \tag{17}$$

Step 2. For columns 1, 2, 3, 6, and 7 crossed by the horizontal lines h_1 and h_4 in Matrix (17), five vertical lines v_1, v_2, v_3, v_6, and v_7 are drawn as shown in Matrix (18).

$$\begin{array}{c} \\ 1 \\ 2 \\ 3 \\ 4 \\ 5 \\ 6 \\ 7 \\ \\ \end{array} \begin{array}{ccccccccccc} 1 & 2 & 3 & 4 & 5 & 6 & 7 & 8 & 9 & 10 & 11 \\ & 1 & 1 & & & & 1 & & & & \\ 1 & & & & 1 & & & & & & 1 \\ & & & & & & & & & 1 & \\ 1 & & 1 & & & 1 & & & & & \\ & & & & 1 & & & 1 & & & \\ 1 & & & 1 & & & & 1 & 1 & 1 & \\ & & 1 & 1 & & 1 & 1 & & 1 & & \\ v_1 & v_2 & v_3 & & & v_6 & v_7 & & & & \end{array} \begin{array}{l} \\ \text{-- } h_1 \\ \\ \\ \text{-- } h_4 \\ \\ \\ \\ \\ \end{array} \tag{18}$$

Step 3. Three horizontal lines, h_2, h_6, and h_7, are drawn through rows 2, 5, and 7, corresponding to the crossed-once elements "1" of Matrix (18). The result of this operation is shown in Matrix (19).

$$\begin{array}{c} \\ 1 \\ 2 \\ 3 \\ 4 \\ 5 \\ 6 \\ 7 \\ \\ \end{array} \begin{array}{ccccccccccc} 1 & 2 & 3 & 4 & 5 & 6 & 7 & 8 & 9 & 10 & 11 \\ & 1 & 1 & & & & 1 & & & & \\ 1 & & & & 1 & & & & & & 1 \\ & & & & & & & & & 1 & \\ 1 & & 1 & & & 1 & & & & & \\ & & & & 1 & & & 1 & & & \\ 1 & & & 1 & & & & 1 & 1 & 1 & \\ & & 1 & 1 & & 1 & 1 & & 1 & & \\ v_1 & v_2 & v_3 & & & v_6 & v_7 & & & & \end{array} \begin{array}{l} \\ \text{-- } h_1 \\ \text{-- } h_2 \\ \\ \text{-- } h_4 \\ \\ \text{-- } h_6 \\ \text{-- } h_7 \\ \\ \end{array} \tag{19}$$

Based on all the crossed-twice elements of matrix (19) a temporary machine cell MC′-1 which includes machines {1, 2, 4, 6, 7} is formed. This temporary machine cell is evaluated by the user. Assume that the user's expertise indicates that machines 2 and 6 cannot be included in MC′-1. Since machines 2 and 6 are excluded, the horizontal lines h_2 and h_6 are ereased from Matrix (19). This leads to the deletion of columns (parts) 1, 4, and 9 from Matrix (19). Note that each of the three parts 1, 4, and 9 is to be manufactured on at least one of the machines 2 and 6 lines h_2 and h_6 in Matrix (19) and at least one of the machines 1, 4, and 7 which form machine cell MC-1 (see Matrix (20) and step 4).

The three parts 1, 4, and 9 are placed on the list of parts to be manufactured in the functional machining facility. Since there are no more crossed-once entries "1," no vertical line is drawn. As a result of the above process, Matrix (20) is obtained.

$$\begin{array}{c|cccccccc|l} & 2 & 3 & 5 & 6 & 7 & 8 & 10 & 11 & \\ \hline 1 & 1 & 1 & & & 1 & & & & h_1 \\ 2 & & & 1 & & & & & 1 & \\ 3 & & & & & & & 1 & & \\ 4 & & 1 & & 1 & & & & & h_4 \\ 5 & & & 1 & & & 1 & & & \\ 6 & & & & & & 1 & 1 & & \\ 7 & & 1 & & 1 & 1 & & & & h_7 \\ \hline & v_2 & v_3 & & v_6 & v_7 & & & & \end{array} \qquad (20)$$

Step 4. The crossed-twice entries "1" of Matrix (20) indicate machine cell MC-1 = { 1, 4, 7 } and part family PF-1 = { 2, 3, 6, 7 }.

Step 5. Matrix (20) is transformed into Matrix (21).

$$A^{(2)} = \begin{array}{c} \\ 2 \\ 3 \\ 5 \\ 6 \end{array} \begin{array}{c} \begin{array}{cccc} 5 & 8 & 10 & 11 \end{array} \\ \begin{bmatrix} 1 & & & 1 \\ & & 1 & \\ 1 & 1 & & \\ & 1 & 1 & \end{bmatrix} \end{array} \qquad (21)$$

Step 6. Set $k = k + 1 = 2$ and go to step 1. The second iteration ($k = 2$) resulted in machine cell MC-2 = {2, 3, 5, 6 } and part family PF-2 = {5, 8, 10, 11 }, as shown in Matrix (22).

$$\begin{array}{c|cccc|l} & 5 & 8 & 10 & 11 & \\ \hline 2 & 1 & & & 1 & h_2 \\ 3 & & & 1 & & h_3 \\ 5 & 1 & 1 & & & h_5 \\ 6 & & 1 & 1 & & h_6 \\ \hline & v_5 & v_8 & v_{10} & v_{11} & \end{array} \qquad (22)$$

The final matrix result generated by the extended cluster identification algorithm is presented in Matrix (23),

		PF-1				PF-2						
		2	3	6	7	5	8	10	11	1	4	9
	1	1	1		1							
MC-1	4		1	1						1		
	7		1	1	1						1	1
	2					1			1	1		
MC-2	3							1				
	5					1	1					
	6						1	1		1	1	1

(23)

Machine cell MC-1 and MC-2 and corresponding part families PF-1 and PF-2 are visible in Matrix (23). The corresponding layout is illustrated in Figure 7.

The functional machining facility in Figure 7 should involve a set of machines which are able to perform operations on parts 1, 4, and 9. However, these machines do not have to be identical to machines 4, 7, 2, and 6 [see Matrix (23)], initially designated for machining these three parts. Apart from the approach for manufacturing parts 1, 4, and 9 which is illustrated in Figure 7, three other alternative approaches are possible:

1. Process plans for parts 1, 4, and 9 might be modified, so that each of these parts could be machined in one of the two existing machine cells MC-1 or MC-2.

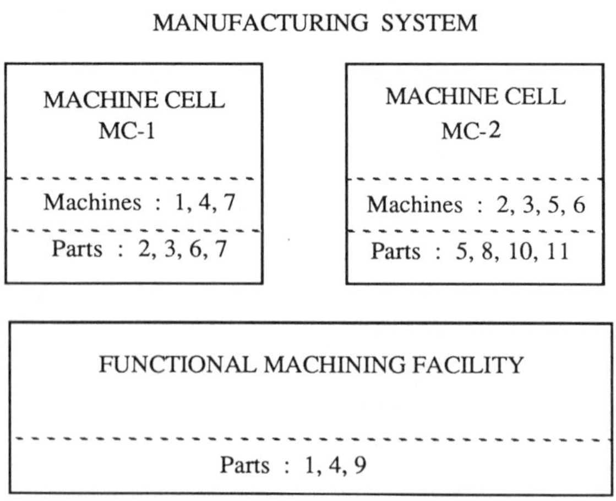

Figure 7. Layout of the manufacturing system from Example 3.

2. Design of parts 1, 4, and 9 might be modified, so that the resulting process plans could fit the existing machine cell MC-1 or MC-2.
3. Parts 1, 4, and 9 might be manufactured in the two existing machine cells MC-1 and MC-2 without any changes in the design or process planning. This approach is applicable as long as the flow of parts among different machine cells (in the above case MC-1 and MC-2) is relatively low.

Finally, if a user is not satisfied with the machine cells and part families generated, he can repeat the entire computational process. Initiating the extended cluster identification algorithm with another machine(s) in step 1 may result in a different configuration of groups of machines and parts.

Another extension of the cluster identification algorithm named the cost analysis algorithm is presented in Kusiak and Chow (1987b). The cost analysis algorithm allows one to solve the group technology problem with known subcontracting costs. The overlapping parts are removed from the incidence matrix in such a way that the total sum of subcontracting costs is minimized.

III. Mathematical Programming Formulation

Most mathematical programming models developed in GT consider a distance measure d_{ij} between parts i and j. The distance measure d_{ij} is a real valued symmetric function obeying the following axioms:

- Reflexivity $d_{ij} = 0$
- Symmetry $d_{ij} = d_{ji}$
- Triangle inequality $d_{iq} = d_{ip} + d_{pq}$

The most commonly used distance measures are the following:

1. Minkowski distance measure (Arthanari and Dodge, 1981)

$$d_{ij} = [\sum_{k=1}^{n} | a_{ik} - a_{jk} |^r]^{1/r}$$

where r is a positive integer and n is the number of parts.

Two special cases of the above measure are widely used:

- Absolute metric measure (for $r = 1$)
- Euclidean metric measure (for $r = 2$)

2. Weighted Minkowski distance measure (Arthanari and Dodge, 1981) is

$$d_{ij} = [\sum_{k=1}^{n} w_k \mid a_{ik} - a_{jk} \mid^r]^{1/r}$$

There are two special cases:

- Weighted absolute metric measure (for $r = 1$)
- Weighted Euclidean metric measure (for $r = 2$)

3. Hamming distance measure (Lee, 1981) is

$$d_{ij} = \sum_{k=1}^{n} \delta (a_{ik}, a_{jk})$$

where

$$\delta (a_{ik}, a_{jk}) = \begin{cases} 1 \text{ if } a_{ik} \neq a_{jk} \\ 0 \text{ otherwise} \end{cases}$$

Distance measures are also referred to as dissimilarity measures. Some similarity measures were discussed in the section on similarity coefficient methods. In this section, the following mathematical programming models are discussed:

- *p*-Median model
- Generalized *p*-median model
- Quadratic programming model

A. THE *p*-MEDIAN MODEL

The *p*-median model is used to group *n* parts into *p* part families. In order to consider the *p*-median model, the following parameters and a variable are defined (Kusiak, 1985):

m = number of machines

n = number of parts

p = number of part families

d_{ij} = distance measure between parts i and j

$$x_{ij} = \begin{cases} 1 \text{ if part } i \text{ belongs to part family } j \\ 0 \text{ otherwise} \end{cases}$$

The objective function of the p-median model is to minimize the total sum of distances between any two parts i and j.

$$\min \sum_{i=1}^{n} \sum_{j=1}^{n} d_{ij} x_{ij} \tag{24}$$

subject to

$$\sum_{j=1}^{n} x_{ij} = 1 \qquad \text{for all } i = 1, \ldots, n \tag{25}$$

$$\sum_{j=1}^{n} x_{jj} = p \tag{26}$$

$$x_{ij} \leqslant x_{jj} \qquad \text{for all } i = 1, \ldots, n \quad j = 1, \ldots, n \tag{27}$$

$$x_{ij} = 0, 1 \qquad \text{for all } i = 1, \ldots, n \quad j = 1, \ldots, n \tag{28}$$

Constraint (25) ensures that each part belongs to exactly one part family. Constraint (26) specifies the required number of part families. Constraint (27) ensures that part i belongs to part family j only when this part family is formed. The last constraint (28) ensures integrality. Note that, in the p-median model, the required number of part families, p, is specified initially. The p-median model (24) through (28) is illustrated in Example 5.

Example 5

Given the machine-part incidence matrix (1), form $p = 2$ part families and their corresponding machine cells. For the machine-part matrix (1), the following matrix of Hamming distances d_{ij} is obtained:

part number

$$[d_{ij}] = \begin{matrix} & 1 & 2 & 3 & 4 & 5 \\ 1 & 0 & 4 & 0 & 4 & 3 \\ 2 & 4 & 0 & 4 & 0 & 1 \\ 3 & 0 & 4 & 0 & 4 & 3 \\ 4 & 4 & 0 & 4 & 0 & 1 \\ 5 & 3 & 1 & 3 & 1 & 0 \end{matrix} \quad \text{part number}$$

Solving the p-median model (24) to (28) for the above matrix and $p = 2$ provides the solution below.

$$x_{11} = 1, x_{31} = 1$$

$$x_{24} = 1, x_{44} = 1, x_{54} = 1$$

Based on the definition of x_{ij}, two part families are formed: PF-1 = {1, 3} and PF-2 = {2, 4, 5}. For the two part families, the corresponding two machine cells are determined from Matrix (1): MC-1 = {2, 4} and MC-2 = {1, 3}.

One can easily notice that the solution to the *p*-median model is identical to that solution obtained using the matrix formulation [see Matrix (2)].

B. GENERALIZED *p*-MEDIAN MODEL

Model (24) to (28) was developed under the assumption that each part i corresponds only to one set of machine operations, known as a process plan. To relax this constraint, Kusiak (1987a) modified model (24) to (28) so that it permits one to consider for each part more than one process plan. In addition, a production cost is associated with each process plan. The objective is to minimize the total sum of distance measures and production costs.

To present the generalized p-median model the following notation is used: F_k is the set of process plans for part number k, $k = 1, \ldots, l$, p the required number of part families, d_{ij} the distance measure between process plans i and j, and c_j the production cost of process plan j.

$$\min \sum_{i=1}^{n} \sum_{j=1}^{n} d_{ij} x_{ij} + \sum_{j=1}^{n} c_j x_{jj} \tag{29}$$

subject to

$$\sum_{i \in F_k} \sum_{j=1}^{n} x_{ij} = 1 \qquad \text{for all } k = 1, \ldots, l \tag{30}$$

$$\sum_{j=1}^{n} x_{jj} \leq p \tag{31}$$

$$x_{ij} \leq x_{jj} \qquad \text{for all } i = 1, \ldots, n;\ j = 1, \ldots, n \tag{32}$$

$$x_{ij} = 0,1 \qquad \text{for all } i = 1, \ldots, n;\ j = 1, \ldots, n \tag{33}$$

Constraint (30) ensures that for each part (a set of process plans) exactly one process plan is selected. Constraint (31) imposes the upper bound on a number of part (process) families. Constraints (32) and (33) correspond to constraints (27) and (28) in the model (24) to (28), respectively. Model (29) to (33) is illustrated in Example 6.

Example 6

Given the machine-part incidence Matrix (34) with the corresponding vector of production costs (35), form $p = 2$ part families and corresponding machine cells using model (29) to (33).

$$
\begin{array}{ccccc}
 & & \text{part number} & & \\
1 & 2 & 3 & 4 & 5 \\
 & & \text{process plan number} & & \\
\overbrace{1\;\,2\;\,3} & \overbrace{4\;\,5} & \overbrace{6\;\,7} & \overbrace{8\;\,9} & \overbrace{10\;\,11}
\end{array}
$$

$$
[a_{ij}] = \left[\begin{array}{ccccccccccc}
 & & 1 & & 1 & 1 & & 1 & 1 & & 1 \\
 & 1 & 1 & 1 & & & 1 & & & & \\
1 & & & 1 & 1 & & & & 1 & 1 & \\
1 & 1 & & & & 1 & 1 & 1 & & 1 &
\end{array}\right]
\begin{array}{l} 1 \\ 2 \\ 3 \\ 4 \end{array}
\begin{array}{l} \\ \text{machine} \\ \text{number} \\ \\ \end{array}
\tag{34}
$$

For simplicity, assume that the vector of production costs

$$c_j = [1, 1, 1, 1, 1, 1, 1, 1, 1, 1, 1,] \tag{35}$$

The Hamming distances d_{ij} are shown in Matrix (36).

$$
\begin{array}{c}
\text{process plan} \\
\begin{array}{ccccccccccc}
1 & 2 & 3 & 4 & 5 & 6 & 7 & 8 & 9 & 10 & 11
\end{array}
\end{array}
$$

$$
[d_{ij}] = \left[\begin{array}{ccccccccccc}
0 & \infty & \infty & 2 & 2 & 2 & 2 & 2 & 2 & 0 & 3 \\
\infty & 0 & \infty & 2 & 0 & 2 & 0 & 2 & 0 & 2 & 3 \\
\infty & \infty & 0 & 2 & 2 & 2 & 2 & 2 & 2 & 2 & 1 \\
2 & 2 & 2 & 0 & \infty & 4 & 2 & 4 & 2 & 2 & 3 \\
2 & 4 & 2 & \infty & 0 & 2 & 0 & 2 & 0 & 2 & 1 \\
2 & 2 & 2 & 4 & 2 & 0 & \infty & 0 & 2 & 2 & 1 \\
2 & 0 & 2 & 2 & 4 & \infty & 0 & 2 & 4 & 2 & 3 \\
2 & 2 & 2 & 4 & 2 & 4 & 2 & 0 & \infty & 2 & 4 \\
2 & 0 & 2 & 2 & 0 & 2 & 4 & \infty & 0 & 2 & 1 \\
0 & 2 & 4 & 2 & 2 & 2 & 2 & 2 & 2 & 0 & \infty \\
3 & 3 & 1 & 3 & 1 & 1 & 3 & 1 & 1 & \infty & 0
\end{array}\right]
\begin{array}{l} 1 \\ 2 \\ 3 \\ 4 \\ 5 \\ 6 \\ 7 \\ 8 \\ 9 \\ 10 \\ 11 \end{array}
\quad
\begin{array}{c} \text{p} \\ \text{r} \\ \text{o} \\ \text{c} \\ \text{e} \\ \text{s} \\ \text{s} \\ \\ \text{p} \\ \text{l} \\ \text{a} \\ \text{n} \end{array}
\tag{36}
$$

All entries in Matrix (36) equal to ∞ denote that, for each F_k, only one process plan is to be selected. Solving the model (29) to (33) for the above data with the LINDO software produces the following solution:

$$x_{27} = 1,\ x_{77} = 1$$

$$x_{59} = 1,\ x_{99} = 1,\ x_{11,9} = 1$$

This solution can be interpreted as follows: PF-1 = {2,7} and PF-2 = {5,9,11}.

For these two part families from Matrix (34), two machine cells are obtained: MC-1 = {2,4} and MC-2 = {1,3}.

The final result is shown in Matrix (37).

	part family PF-1		part family PF-2				
	2	7	5	9	11		
	1	1				2	machine
	1	1				4	cell MC-1
			1	1	1	1	machine
			1	1		3	cell MC-2

(37)

C. QUADRATIC PROGRAMMING MODEL

Kusiak *et al.* (1986) developed a formulation of the GT problem that allows one to deal with the restricted number of clusters and cluster sizes. The formulation can be expressed in the form of the following 0–1 quadratic programming model:

$$\min \sum_{i=1}^{n-1} \sum_{j=i+1}^{n} \sum_{l=1}^{p} d_{ij}\, x_{il}\, x_{jl} \tag{38}$$

subject to

$$\sum_{j=1}^{p} x_{ij} = 1 \qquad \text{for all } i = 1, \ldots, n \tag{39}$$

$$\sum_{i=1}^{n} x_{ij} = m_j \qquad \text{for all } j = 1, \ldots, p \tag{40}$$

$$x_{ij} = 0,1 \qquad \text{for all } i = 1, \ldots, n; \quad j = 1, \ldots, p \tag{41}$$

Constraint (39) ensures that each part belongs to exactly one part family. Constraint (40) imposes that part family j contains exactly m_j parts. Constraint (41) ensures integrality. The model was solved by an eigenvector-based algorithm (Kusiak *et al.*, 1986).

IV. Graph Formulation

In the graph formulation, the incidence matrix $[a_{ij}]$ is represented by a graph. The following three type of graphs can be used: bipartite, transition, and boundary.

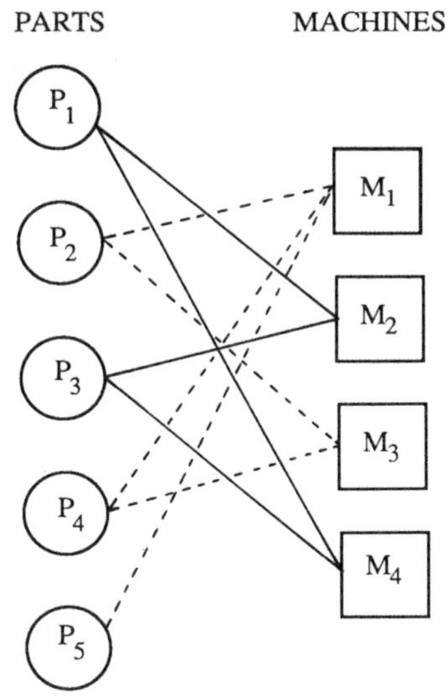

Figure 8. Bipartite graph corresponding to Matrix (1).

A. BIPARTITE GRAPH

In the bipartite graph, one set of nodes represents parts and the other machines. In order to discuss the application of bipartite graphs to GT, consider the graph in Figure 8 representing the incidence matrix (1). It can be seen that the bipartite graph in Figure 8 decomposes into two disjoint bipartite graphs shown in Figure 9.

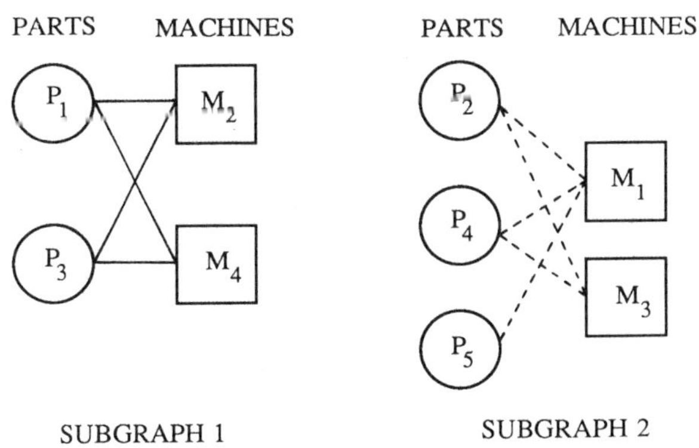

Figure 9. Two disjoint bipartite graphs for the graph in Figure 8.

Based on the graphs in Figure 9, the following two part families are formed: PF-1 = {1, 3} and PF-2 = {2, 4, 5}. The corresponding two machine cells are MC-1 = {2, 4} and MC-2 = {1, 3}.

It is easily seen that the solution for the bipartite graph is identical to the solution obtained using the previous two approaches. Readers interested in representation with boundary and transition graphs may refer to Kusiak and Chow (1988).

V. KBGT: A Knowledge-Based System for Group Technology

In this section, a formulation of the group technology problem that generalizes all the models discussed in the previous sections is presented.

Rather than the binary matrix $[a_{ij}]$, the matrix $[t_{ij}]$ where $t_{ij} \geqslant 0$ is the processing time of part number j on machine number i is considered. The formulation involves also some constraints, which are typical to flexible manufacturing environments and have not been considered in the previous sections.

The group technology problem in flexible manufacturing systems can be loosely formulated as follows (Kusiak, 1987b):

> Determine machine cells and part families with a minimum number of parts visiting more than one machine cell, and select a suitable material handling carrier with the minimum corresponding cost subject to the following constraints:
>
> Constraint C1: Processing time available at each machine is not exceeded.
>
> Constraint C2: Upper limit on the frequency of trips of material handling carriers for each machine cell is not exceeded.
>
> Constraint C3: Number of machines in each machine cell does not exceed its upper limit or alternatively the dimension (e.g., the length) of each machine cell is not exceeded.

The above formulation of the GT problem is not only computationally complex, but also involves constraints that are difficult to handle by any algorithm alone. Therefore, to solve the above problem, a knowledge-based system has been developed.

A. STRUCTURE OF THE KNOWLEDGE-BASED SYSTEM

In this section, a tandem knowledge-based system for group technology (KBGT) is considered. All models and algorithms discussed in the previous sections can be incorporated into its model and algorithm base. A group tech-

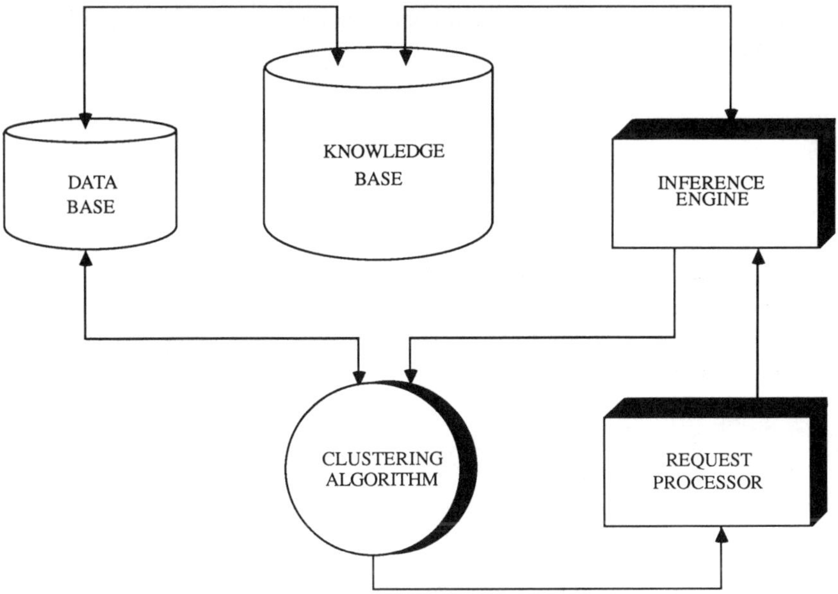

Figure 10. Structure of the knowledge-based system.

nology problem that could be modeled by any of the models discussed would be solved by a suitable algorithm. To simplify further considerations, only one clustering algorithm presented in Section D is embedded into the KBGT. The algorithm is suitable for solving the group technology problem in flexible manufacturing systems as well as its special cases.

The KBGT considered has the structure shown in Figure 10. The KBGT consists of five components:

- Database
- Knowledge base
- Inference engine
- Request processor
- Clustering algorithm

These components are discussed in detail later in this section.

The KBGT has been implemented in Common LISP on a SPERRY-MICRO-IT (an IBM-PC compatible). LISP, as programming language for KBGT, has been selected for three reasons:

- It facilitates implementation of the declarative knowledge.
- It facilitates implementation of the procedural knowledge (the clustering algorithm).

• It provides flexibility to define and implement the interaction between the algorithm and the knowledge-based subsystem.

For the source LISP code of KBGT, see Kusiak (1988).

1. Input Data

The input data required by KBGT fall into two categories: machine data and part data. In addition, depending on the characteristics of the manufacturing system, the following optional data can be provided:

• Maximum number of machines in a machine cell

• Maximum frequency of trips which can be handled by a material handling carrier (for example, robot or automated guided vehicle, AGV)

2. Grouping Process

Before beginning the grouping process, KBGT constructs a machine-part incidence matrix based on the data provided by the user.

Next, the KBGT initializes in the database objects representing facts known about the manufacturing system considered. Then the system forms machine cells and the corresponding part families. Each machine cell is formed by including one machine at a time. A machine is first analyzed by the KBS for possible inclusion in the machine cell. For example, a bottleneck machine, that is, the machine that processes parts visiting more than one machine cell, is not included.

Each time a machine cell has been formed, the KBS checks whether any of the constraints C1 to C3 have been violated and removes from the machine-part incidence matrix all parts violating the constraints.

For a machine cell, that has been formed and analyzed by the KBS, the corresponding machines and parts forming a part family are removed from the machine-part incidence matrix. The system does not backtrack in the grouping process, that is, once a machine cell is formed, the machines included in the cell are not considered for future machine cells.

3. Output Data

At the end of the grouping process, KBGT prints the following data:

• *Machine cells formed:* The machine cells formed are listed in the order they have been generated. For each machine cell the following information is provided: (1) machine cell number, (2) list of machines in a machine cell, (3) part family number, and (4) list of part numbers in a part family.

• *Part waiting list:* This list includes parts that were placed on the waiting list due to either overlapping of parts in such a way that prevents grouping or including them in a machine cell would violate one or more constraints.

• *List of machines not used:* The list of machines with all parts removed during the grouping process.

• *List of bottleneck machines:* The list of machines processing parts that need to be processed on machines belonging to more than one machine cell. These machines should be given special consideration while determining the layout. Preferably, each of these machines should be located adjacent to the machine cell that processes the same parts.

• *Maximum number of machines in a cell:* This number indicates the maximum number of machines in a machine cell. It has an impact on the machine cells formed, namely, if it is too small it might result in the removal of too many parts. If this number is not provided by the user, the system groups the machines based on their similarities only.

For any two 0–1 vectors

$$m_i = [a_{i1}, \ldots a_{ik}, \ldots, a_{in}]$$

$$m_j = [a_{j1}, \ldots a_{jk}, \ldots, a_{jn}]$$

define a similarity measure s_{ij}

$$s_{ij} = \sum_{k=1}^{n} \delta\,(a_{ik}, a_{jk})$$

where

$$\delta(a_{ij}, a_{jk}) = \begin{cases} 1 \text{ if element } a_{ik} = a_{jk} \\ 0 \text{ otherwise} \end{cases}$$

$$n = \text{number of parts}$$

In particular, vector m_i may represent all parts included in machine cell MC-k, and vector m_j may represent parts corresponding to the machine to be selected (see step 2 in the clustering algorithm presented in Section D). In this case, the distance s_{ij} is regarded as a distance between machine cell MC-k and machine m_j.

B. DATABASE

The global database contains information about the current problem represented in a form of objects and frames. It is a nonmonotonic database, since the objects are modified by the clustering algorithm and the knowledge-based subsystem (KBS).

The content of the database is either provided by the user as input data, or generated by the system. A list of objects and frames in the data base is as follows:

• *Machine frame* contains information regarding each machine and has the following format:

```
(machine#i (( parts ((p#1  p-time) . . .
               (p#j  p-time) . . . ))
               (max-process-time  x)
               (multiple  y)))
```

where machine #*i* is machine number *i* (frame identifier); *p*-time is the time required to process part number *j* on machine number *i* (*p*-time is equal to 0, if processing time is not available); max-process-time (optional) is the maximum processing time available on machine m#*i*, i.e., the capacity of machine number *i*; and multiple (optional) is the number of the identical machines available.

• *Part frame* contains information regarding each part and has the following format:

```
(part#j  ((primary-pp  (m#1 . . . m#i . . . ))
          (fr  y)
          (fa  z)))
```

where part #*j* is part number *j* (frame identifier); primary-*pp* is the primary process plan for part number *j*; *fr* is the frequency of trips required by a robot to handle part number *j*; and *fa* is the frequency of trips required by an AGV to handle part number *j*.

• *Matrix-t (machine-part incidence matrix)* is constructed by the system based on the input data provided in the following format:

```
((m#1  ((p#1 p-time) (p#2 p-time) . . . ))
 (m#i . . .                             )
      ⋮                                  )
```

In cases when processing times are not available, by default the matrix is a 0–1 incidence matrix, i.e., *p*-time is 0.

• *Current machine* is a machine that the system has selected for possible inclusion in the machine cell currently being formed.

• *List of candidate machines* is a list of candidate machines to be included in the machine cell currently being formed.

• *List of temporary candidate machines* is a list of all machines such that the parts that are processed on the current machine are also processed on one or more of these machines. Moreover, these machines are not on the list of candidate machines.

• *Part waiting list* is a list of parts that have been removed from the machine-part incidence matrix.

• *List of bottleneck machines* is a list of all bottleneck machines.

• *List of temporary bottleneck machines* is a list of machines that are considered to be temporary bottleneck machines. These machines may become nonbottleneck machines after some parts have been removed from the machine-part incidence matrix.

• *List of machines not used* (same as discussed earlier in the subsection on the output data from KBGT).

• *MC-k (machine cell k)* is a list of machine numbers in the current machine cell.

• *PF-k (part family k)* is a list of part numbers in the current part family.

C. THE KNOWLEDGE-BASED SUBSYSTEM

As illustrated in Figure 10, the knowledge-based subsystem consists of three components: knowledge base, inference engine, and request processor.

1. Knowledge Base

The knowledge base in KBGT contains production rules, which have been acquired from three experts in group technology and the literature. In the implementation of KBGT discussed in this chapter, the knowledge base consists of three classes of production rules:

- Preprocessing rules
- Current machine rules
- Machine cell rules

The preprocessing rules deal with the initialization of objects in the database that are not provided by the user. The current machine rules check the appropriateness of a current machine to the machine cell being formed, for example, whether the current machine is a bottleneck machine. The machine cell rules deal with each machine cell which has been formed. Machine cell rules check for violation of constraints and remove parts violating them. Structuring of rules into separate classes has two advantages. First, the search for applicable rules is more efficient since the inference engine attempts to fire only rules that are relevant to the current context. Second, the modularity of a rule base makes it more understandable, and easy for modification.

Each rule has the following format:

Rule number
IF conditions

THEN actions

The rule number is used for identification by the inference engine. The most significant digit represents a class, and the other two digits represent a rule number in a class. Each condition in a rule has one of the following three forms:

1. A straightforward checking of values in the data base
2. Procedure calls to calculate the values required
3. A combination of (1) and (2)

An example of (1) is comparing the size of the current MC-*k* with the maximum number of machines allowed in the machine cell. An example of (2) is a call of the procedure calculating the similarity between a current machine and a machine cell.

Sample production rules that have been implemented in KBGT are listed below:

Rule 102 (preprocessing rule)

IF the intercellular movement (icm) level is not specified

AND the total number of machines is greater than 50

THEN set icm to 3%.

Note that the intercellular movement level is defined as $icm = n_1/n \; 100\%$, where n_1 is the number of overlapping parts and n is the total number of parts considered.

Alternatively, the value of *icm* can be set by a system user.

Rule 103 (preprocessing rule)

IF the maximum number of machines in a machine cell is specified

THEN remove from the machine-part incidence matrix all parts that require more machines than the maximum number of machines in the machine cell

AND place them on the part waiting list.

Rule 201 (current machine rule)

IF no machine has been included in MC-k

AND the number of temporary candidate machines plus the current machine is greater than the maximum number of machines in a machine cell

THEN add the current machine number to the list of temporary bottleneck machines

AND go to step 1 of the clustering algorithm (see Section D).

Rule 202 (current machine rule)

IF the maximum number of machines in a machine cell is not specified

AND the similarity between the current machine and MC-k is less than similarity of the next machine to be selected as current machine

AND the number of parts, that are processed by the current machine and machines in MC-k, is less than or equal to one indicated by the intercellular movement (icm) level

THEN place the parts mentioned above in the part waiting list

AND set the list of temporary candidate machines to empty

AND set the list of candidate machines to empty

AND set the current machine to null

AND go to step 6 of the clustering algorithm (see Section D).

Rule 203 (current machine rule)

IF the total number of machines in MC-k and machines in the list of temporary candidate machines and machines in the list of candidate machines is greater than the maximum number of machines in a machine cell

AND the number of parts, which are processed by the current machine and machines in MC-k, is less than or equal to one indicated by the intercellular movement (icm) level

THEN place the parts mentioned above in the part waiting list

AND set the list of temporary candidate machines to empty

AND set the current machine to null

AND go to step 5 of the clustering algorithm (see Section D).

Rule 302 (machine cell rule)

IF there are machines where constraint C1 is violated

THEN remove parts from the machines violating constraint C1

AND place the removed parts on the part waiting list.

Rule 303 (machine cell rule)

IF constraint C2 is violated for a robot or AGV

THEN select a robot or AGV such that C2 is not violated.

2. Inference Engine

One of the greatest advantages of the tandem system architecture is the simplicity of the inference engine. The inference engine in KBS employs a forward-chaining control strategy. In a given class of rules, it attempts to fire all the rules which are related to the context considered. If a rule is triggered, that is, the conditions are true, the actions of the triggered rule are carried out. However, some rules, for example Rule 201, Rule 202, and Rule 203, stop the search of the knowledge base and send a message to the algorithm.

3. Request Processor

The request processor facilitates the interaction between the algorithm and KBS. Based on each request of the algorithm, the request processor calls the inference engine and selects a suitable class of rules to be searched by the inference engine.

D. CLUSTERING ALGORITHM

The clustering algorithm presented is an extension of the cluster identification algorithm discussed in the previous section.

Algorithm

Step 0. Set iteration number $k = 1$. Construct machine-part incidence matrix. Send a request to KBS for preprocessing.

Step 1. Select a machine (row of machine-part incidence matrix) that processes the maximum number of parts and is not in the list of temporary bottleneck machines. Place the selected machine in the list of candidate machines.

Step 2. From the list of candidate machines, select a machine which is the most similar to machine cell MC-k. If machine cell MC-k is empty, choose the machine selected in step 1. Draw a horizontal line h_i, where i is the selected machine number.

Step 3. For each entry crossed once by the horizontal line h_i draw a vertical line v_j. Parts indicated by the vertical lines are potential candidates for part family PF-k. For each entry $t_{ij} > 0$ crossed by a vertical line v_j, add the corresponding machines, which are not in the list of candidate machines to the list of temporary candidate machines. Remove the current machines from the list of candidate machines.

Step 4. KBS analyzes the current machine selected and takes one of the following two actions:
- Go to step 5 (include the current machine in MC-*k*)
- Go to step 1 (do not include it)

Step 5. Add the machine considered to machine cell MC-*k*. Add the corresponding part numbers to part family PF-*k*. If the list of candidate machines is empty, go to step 6; otherwise, go to step 2.

Step 6. KBS analyzes machine cell MC-*k* for violations of constraints C1-C3 and attempts to satisfy the constraints. Remove machine cell MC-*k* and part family PF-*k* from the machine-part incidence matrix.

Step 7. If the machine-part incidence matrix is not empty, increment *k* by 1 and go to step 1; otherwise, STOP.

E. ILLUSTRATIVE EXAMPLE

Given the machine-part incidence matrix (42), vector *fa* (frequency of AGV trips required for handling each part), vector *fr* (frequency of robot trips required to handle each part), max *fa* = 40 (maximum frequency of trips that can be handled by an AGV), max *fr* = 100 (maximum frequency of trips that can be handled by a robot), and vector *T* [the column outside of Matrix (42)], solve the group technology problem. The maximum number of machines in a machine cell is 3.

	01	02	03	04	05	06	07	08	09	10	11	12	
fa	11	30	2.5	-	6	10	-	6	7	15	18	14	max-fa (40)
fr	11	30	5	3	6	16	10	12	7	-	36	28	max-fr (100)
PART NUMBER / MACHINE NUMBER	01	02	03	04	05	06	07	08	09	10	11	12	T
1		4		21				8					40
2	26		5			10							40
3			20		10					22		8	40
4		35					2	6					50
5	5					6			25				50
6		16		10			3				1		60
7					1					7		7	20

(42)

Iteration 1

Step 0. Iteration number is set to $k = 1$.
The machine-part incidence matrix is constructed from the input data presented in Matrix (42). A request is sent to KBS for preprocessing. KBS initializes the following lists to be empty: MC-k, PF-k, candidate machines, temporary candidate machines, bottleneck machines, temporary bottleneck machines, current machine, and machines not used list.

Step 1. Machine 3 is selected since it processes the maximum number of parts and is not in the list of temporary bottleneck machines. It is placed in the list of candidate machines.

Step 2. Machine 3 is selected from the list of candidate machines. A horizontal line h_3 is drawn as shown in Matrix (43).

fa [11 30 2.5 - 6 10 - 6 7 15 18 14] max-fa (40)

fr [11 30 5 3 6 15 10 12 7 - 36 28] max-fr (100)

			v3		v5					v10		v12		
MACHINE NUMBER \ PART NUMBER	01	02	03	04	05	06	07	08	09	10	11	12		
1		4		21				8					40	
2	26		5			10							40	
3			20		10					22		8	40	h_3
4		35					2	6					50	
5	5					6			25				50	
6		16		10			3				1		60	
7					1					7		7	20	

(43)

Step 3. Vertical lines v_3, v_5, v_{10}, and v_{12} are drawn [see Matrix (43)]. Machines 2 and 7 are added to the list of temporary candidate machines. Machine 3 is removed from the list of candidate machines.

Step 4. Since the total number of machines in MC-k, machines in the list of candidate machines, machines in the list of temporary candidate machines, and the current machine equals 3, Constraint

C3 is not violated. No current machine rule is fired. Go to step 5.

Step 5. Machine 3 is added to MC-k and parts 3, 5, 10, and 12 are included in PF-k. Since the list of candidate machines is not empty, go to step 2.

Step 2. Machine 7 is selected as the most similar machine to the cell MC-k. A horizontal line h_7 is drawn.

Step 3. Vertical lines v_5, v_{10}, and v_{12} have already been drawn. Machine 7 is removed from the list of candidate machines.

Step 4. Since the total number of machines in MC-k, machines in the list of candidate machines, machines in the list of temporary candidate machines, and the current machine equals 3, constraint C3 is not violated. Go step 5.

Step 5. Machine 7 is added to MC-k. Since the list of candidate machines is not empty, go to step 2.

Step 2. Machine 2 is selected from the list of candidate machines. A horizontal line h_2 is drawn [see Matrix (44)].

fa [11 30 2.5 - 6 10 - 6 7 15 18 14] max-fa (40)

fr [11 30 5 3 6 15 10 12 7 - 36 28] max-fr (100)

	v1		v3		v5	v6				v10		v12		
PART NUMBER	01	02	03	04	05	06	07	08	09	10	11	12		
MACHINE NUMBER 1		4		21				8					40	
2	26		5			10							40	h_2
3			20		10					22		8	40	h_3
4		35					2	6					50	
5	5					6			25				50	
6		16		10			3				1		60	
7					1					7		7	20	h_7

(44)

Step 3. Vertical lines v_1 and v_6 are drawn [see Matrix (44)]. Machine 5 is added to the list of temporary candidate machines. Machine 2 is removed from the list of candidate machines.

Step 4. Since the total number of machines in MC-k and machines in the list of candidate machines and machines in the list of temporary candidate machines and the current machine equals 4, Rule 203 is fired:

- Part 3 is placed in the part waiting list.
- The list of temporary candidate machines is set to empty.
- The current machine is set to empty.

Now Constraint C3 is no longer violated. Go to step 5.

Step 5. Since the current machine is empty, no element is added to MC-k and PF-k. Since the list of candidate machines is empty, go to step 6.

Step 6. Constraint C2 is violated, because of part 10 which cannot be handled by an AGV. Therefore, Rule 303 is fired and a robot is selected as the material handling alternative. Machines 3 and 7 and parts 5, 10, and 12 are removed from the machine-part incidence matrix [see Matrix (45)]. Iteration number k is incremented by 1.

$$
\begin{array}{lll}
\text{fa} & \begin{bmatrix} 11 & 30 & - & 10 & - & 6 & 7 & 18 \end{bmatrix} & \text{max-fa (40)} \\
\text{fr} & \begin{bmatrix} 11 & 30 & 3 & 15 & 10 & 12 & 7 & 36 \end{bmatrix} & \text{max-fr (100)}
\end{array}
$$

$$
\begin{array}{ll|cccccccc|l}
\text{PART NUMBER} & & 01 & 02 & 04 & 06 & 07 & 08 & 09 & 11 & \\
\hline
 & 1 & & 4 & 21 & & & 8 & & & 40 \\
 & 2 & 26 & & & 10 & & & & & 40 \\
\text{MACHINE NUMBER} & 4 & & 35 & & & 2 & 6 & & & 50 \\
 & 5 & 5 & & & 6 & & & 25 & & 50 \\
 & 6 & & & 10 & & 3 & & & 1 & 60
\end{array}
\qquad (45)
$$

In two more iterations of the clustering algorithm, the result shown in Matrix (46) and the corresponding layout in Figure 11 are produced.

$$
\begin{array}{c|cccccccccccc}
 & 1 & 6 & 9 & 5 & 10 & 12 & 2 & 4 & 7 & 8 & 11 & 3 \\
\hline
2 & 26 & 10 & & & & & & & & & & 5 \\
5 & 5 & 6 & 25 & & & & & & & & & \\
3 & & & & 10 & 22 & 8 & & & & & & 20 \\
7 & & & & 1 & 7 & 7 & & & & & & \\
1 & & & & & & & 4 & 21 & & 8 & & \\
4 & & & & & & & 35 & & 2 & 6 & & \\
6 & & & & & & & 16 & 10 & 3 & & 18 &
\end{array}
\qquad (46)
$$

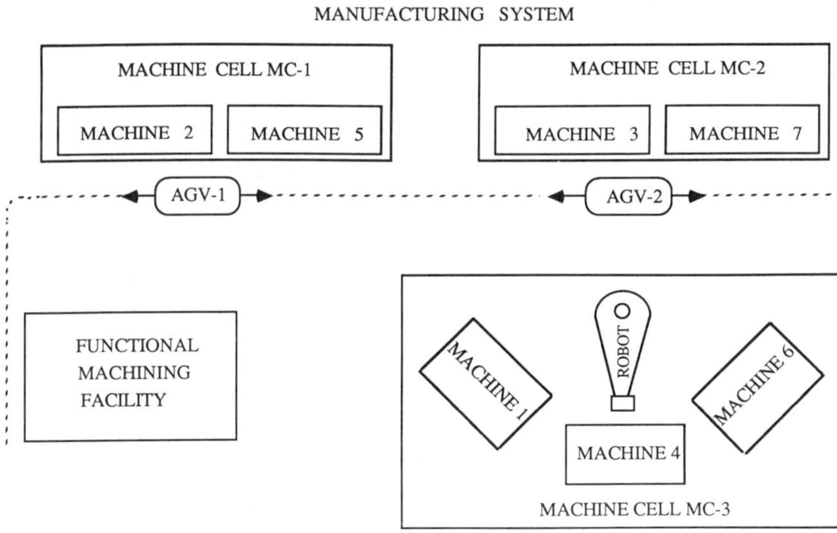

Figure 11. Layout corresponding to the solution of the group technology problem represented in Matrix (46).

As shown in Matrix (46) and Figure 11, three machine cells MC-1 = {2, 5}, MC-2 = {3, 7}, MC-3 = {1, 4, 6} and the three corresponding part families PF-1 = {1, 6, 9}, PF-2 = {5, 10, 12}, PF-3 = {2, 4, 7, 8, 11} have been generated. Part number 3 has been assigned to a functional manufacturing facility. Two AGVs and a handling robot tend the three machine cells.

F. APPLICATION OF KBGT

To illustrate and test performance of the KBGT, a number of problems have been considered. The first is a generalized group technology problem represented by Matrix (46). The KBGT input representation of Matrix (46) is shown in Figure 12.

The output from KBGT for Matrix (42) is shown in Figure 13.

As shown in Figure 13, three machine cells and three part families have been generated. MC-1 is served by an AGV, MC-2 is served by a robot, and MC-3 can be served by a robot or an AGV. The overlapping part 3 is placed on the part waiting list. The computation was performed for the maximum cell size equal 3.

The second problem is a special case of the generalized GT problem. It is based on 0–1 machine part incidence matrix shown in Figure 14.

```
(setq  machine-db  ' ((m1 ((parts ((p2  4) (p4  21) (p8  8)))
                          (max-process-time  40)))

                      (m2 ((parts((p1  26) (p3  5) (p6  10)))
                          (max-process-time  40)))

                      (m3 ((parts((p3  20) (p5  10) (p10  22) (p12  8)))
                          (max-process-time  40)))

                      (m4 ((parts ((p2  35) (p7  2) (p8  6)))
                          (max-process-time  50)))

                      (m5 ((parts ((p1  5) (p6  6) (p9  25)))
                          (max-process-time  50)))

                      (m6 ((parts ((p2  16) (p4  10) (p7  3)(p11  18)))
                          (max-process-time  60)))

                      (m7 ((parts ((p5  1) (p10  7) (p12  7)))
                          (max-process-time  20)))
                                                                  ))

(setq part-db  '     ((p1 ((primary-pp (m2  m5))
                        (fr  11)  (fa  11)))

                     ((p2 ((primary-pp (m1  m4  m6))
                        (fr  30)  (fa  30 )))

                     ((p3 ((primary-pp (m2  m3))
                           (fr  5)  (fa  2.5)))

                     ((p4 ((primary-pp (m1  m6))
                           (fr  3)  (fa  0)))

                     ((p5 ((primary-pp (m3  m7))
                           (fr  6)  (fa  6)))

                     ((p6 ((primary-pp (m2  m5))
                           (fr  15)  (fa  10)))

                     ((p7 ((primary-pp (m4  m6))
                           (fr  10)  (fa  0)))

                     ((p8 ((primary-pp (m1 m4))
                           (fr  12)  (fa  6)))

                     ((p9 ((primary-pp (m5))
                           (fr  7)  (fa  7)))

                     ((p10 ((primary-pp (m3  m7))
                           (fr  0)  (fa  15)))

                     ((p11 ((primary-pp (m6))
                           (fr  36)  (fa  18)))

                     ((p12 ((primary-pp (m3  m7))
                                                  ))
(setq  max-mc-k-size  0)
(setq  max-fr  100)
(setq  max-fa  40)
```

Figure 12. Input of Matrix (42) in KBGT format.

```
( (MACHINE-CELL  1)     M3  M7)
( (PART-FAMILY   1)     P5  P10  P12)
(M-H-S-ALTERNATIVE      (AGV))
        +++++++++++++++++
( (MACHINE-CELL  2)     M6  M1  M4)
( (PART-FAMILY   2)     P2  P4  P7  P11  P8)
(M-H-S-ALTERNATIVE      (ROBOT))
        +++++++++++++++++
( (MACHINE-CELL  3)     M5  M2)
( (PART-FAMILY   3)     P1  P6  P9)
(M-H-S-ALTERNATIVE      (ROBOT OR AGV))
        +++++++++++++++++
---------------------------------------------------
(PART-WAITING-LIST===========>   (P3) )
(MACHINES-NOT-USED==========>     0 )
(BOTTLENECK-MACHINES========>     0 )
---------------------------------------------------
(MAXIMUM-MACHINE-CELL-SIZE-USED   3)
```

Figure 13. KBGT output generated for Matrix (42).

MACHINE NUMBER / PART NUMBER	01	02	03	04	05	06	07	08	09	10	11	12	13	14	15	16	17	18	19	20	21	22	23	24	25	26	27	28	29	30	31	32	33	34	35	36	37	38	39	40	41	42	43
1																																					1					1	
2		1								1																		1				1					1	1		1		1	
3							1										1																	1	1	1							
4					1				1					1					1		1		1						1														
5					1			1	1					1	1	1			1		1		1						1				1								1		1
6	1	1				1	1	1				1	1	1			1		1				1									1	1	1			1		1	1		1	1
7	1												1												1																		
8	1	1	1					1	1		1	1			1				1	1	1		1	1			1	1			1						1	1			1		1
9		1		1						1								1										1				1					1	1		1		1	
10	1											1	1												1	1					1								1				
11			1						1											1				1			1			1													
12											1											1		1			1			1													
13			1																					1																			
14		1				1											1																		1								
15					1									1					1		1												1								1		1
16		1					1			1								1														1					1	1				1	

Figure 14. Machine-part incidence matrix (Burbidge, 1973).

```
( (MACHINE-CELL  1)   M5  M4)
( (PART-FAMILY   1)   P5  P8  P14  P15  P16  P19
                      P21  P23  P29  P33  P41
                      P43)
        +++++++++++++++++
( (MACHINE-CELL  2)   M9  M2  M16  M1  M M3)
( (PART-FAMILY   2)   P2  P4  P10  P18  P28  P32
                      P37  P38  P40  P6  P7  P17
                      P34  P35  P36  P42)
        +++++++++++++++++
( (MACHINE-CELL  3)   M10  M7)
( (PART-FAMILY   3)   P1  P12  P13  P25  P26  P31
                      P39)
        +++++++++++++++++
( (MACHINE-CELL  4)   M12  M11  M13)
( (PART-FAMILY   4)   P11  22  P24  P27  P30  P3
                      P20)
        +++++++++++++++++
---------------------------------------------------
(PART-WAITING-LIST===========>   (P9) )
(MACHINES-NOT-USED===========>     0 )
(BOTTLENECK-MACHINES======> M8    M6 )
---------------------------------------------------
(MAXIMUM-MACHINE-CELL-SIZE-USED   6 )
```

Figure 15. KBGT output for the machine-part incidence matrix in Figure 14.

TABLE 1
Solutions of Four Group Technology Problems

Problems			Solution					
Problem reference	Number of machines	Number of parts	Solution method	Maximum machine cell size	Number of machine cells and part families	Number of overlapping parts	Bottleneck machines	KBGT CPU time
			ROC2 algorithm[1]					
			• solution 1	n/a	—	—	2	
King and			• solution 2	n/a	4	3	2	
Nakornchai (1982)	16	43	KBGT	6	4	1	2	4 s
			SLCA	n/a	2	5	0	
Seifoddini (1986)	5	12	KBGT	3	2	4	0	1 s
			Kumar and Vannelli algorithm					
			• solution 1	n/a	2	4	0	
Kumar and Vannelli			• solution 2	n/a	2	5	0	
(1987)	30	41	• solution 3	n/a	3	6	0	
			KBGT[2]					
			• solution 1	14	3	7	0	5 s
			• solution 2	n/s	3	6	0	7 s
			ZODIAC	n/a	10	33	0	
			KBGT					
			• solution 1	11	5	23	0	23 s
Chandrasekharan and			• solution 2	10	6	32	0	24 s
Rajagopalan (1987)	40	100	• solution 3	9	7	31	0	40 s
			• solution 4	n/s	9	32	0	50 s

n/a not applicable.
n/s not specified.
[1]Solution 1 has been generated by the ROC2 algorithm.
[2]Subcontracting the overlapping part 38 in solution 2 makes machine 24 redundant.

For the incidence matrix in Figure 14, the KBGT provides the solution in Figure 15.

To date, a large number of clustering algorithms have been developed, mostly for solving the 0–1 group technology problem, and only a few of them have been tested.

To evaluate the performance of KBGT, consider four matrix formulations of the GT problem in Table 1. The solutions obtained are of better quality than ones generated by the four algorithms considered. The computational time complexity of the heuristic clustering algorithms available in the literature is high, for example $O(m^2n + n^2m)$, where m is the number of rows and n is the number of columns in a machine-part incidence matrix (see Kusiak, 1985). The algorithm presented in this chapter has the computational time complexity slightly higher than $O(mn)$. The CPU time reported in Table 1 is for a SPERRY-MICRO-IT (an IBM-PC compatible). Note that some of the traditional algorithms listed in Table 1 required human interaction, while KBGT does not.

References

Arthanari, T. S., and Dodge, Y. (1981). "Mathematical Programming in Statistics." Wiley, New York.

Askin, R., and Subramanian, S. (1987). A cost-based heuristic for group technology configuration. *Int. J. Prod. Res.* **25**(1), 101–114.

Bhat, M. V., and Haupt, A. (1976). An efficient clustering algorithm. *IEEE Trans. Syst., Man Cybernet.* **SMC-6**(1), 61–64.

Burbidge, J. L. (1971). Production flow analysis. *Prod. Eng.* April/May, pp. 139–152.

Burbidge, J. L. (1973). "Production Flow Analysis on the Computer," 3rd Annu. Conf. Institute of Production Engineers, Group Technology Division.

Chan, H. M., and Milner, D. A. (1982). Direct clustering algorithm for group formation in cellular manufacturing. *J. Manuf. Syst.* **1**(1), 65–74.

Chandrasekharan, M. P., and Rajagopalan, R. (1986). MODROC: An extension of rank order clustering for group technology. *Int. J. Prod. Res.* **24**(5), 1221–1233.

Chandrasekharan, M. P., and Rajagopalan, R. (1987). ZODIAC—an algorithm for concurrent formation of part-families and machine-cells. *Int. J. Prod. Res.* **25**(6), 835–850.

Cheng, Y., and Fu, K. S. (1985). Conceptual clustering in knowledge organization. *IEEE Trans. Pattern Anal. Mach. Intell.* **PAMI-7**(5), 592–598.

De Witte, J. (1980). The use of similarity coefficients in production flow analysis. *Int. J. Prod. Res.* **18**(4), 503–514.

Everitt, B. (1980). "Cluster Analysis." Halsted Press, New York.

Ingram, F. B. (1982). Group technology. *Prod. Inventory Manage. J., 4th Q.*

Iri, M. (1968). On the synthesis of the loop cut set matrices and the related problem. *RAAG Mem.* **4,** 376–410.

King, J. R. (1980). Machine-component group formation in production flow analysis: An approach using a rank order clustering algorithm. *Int. J. Prod. Res.*, **18**(2), 213–232.

King, J. R., and Nakornchai, V. (1982). Machine-component group formation in group technology: Review and extensions. *Int. J. Prod. Res.* **20**(2), 117–133.

Klastorin, T. D. (1982). The p-median problem for cluster analysis: A comparison test using the mixture model approach. *Manage. Sci.* **31**(1), 1134–1146.

Kumar, K. R., and Vannelli, A. (1987). Grouping of parts and components in flexible manufacturing systems. *Int. J. Prod. Res.* **25**(12), 1715–1728.

Kumar, K. R., Kusiak, A., and Vannelli, A. (1986). Grouping parts and components in flexible manufacturing systems. *Eur. J. Oper. Res.* **24**(3), 387–397.

Kusiak, A., (1985). The part families problem in flexible manufacturing systems. *Ann. Oper. Res.* **3,** 279–300.

Kusiak, A. (1987a). The generalized group technology concept. *Int. J. Prod. Res.* **25**(4), 561–569.

Kusiak, A. (1987b). Artificial intelligence and operations research in the flexible manufacturing systems. *Inf. Process. Oper. Res. (INFOR)* **25**(1), 2–12.

Kusiak, A. (1988). Knowledge-based group technology. In "Expert Systems: Strategies and Solutions in Manufacturing Design and Planning" (A. Kusiak, ed.), pp. 259–303. Society of Manufacturing Engineers, Dearborn, Michigan.

Kusiak, A. (1990). "Intelligent Manufacturing Systems," Prentice Hall, Englewood Cliffs, New Jersey.

Kusiak, A., and Chow, W. S. (1987a). Efficient solving of the group technology problem. *J. Manuf. Syst.* **6**(2), 117–124.

Kusiak, A., and Chow, W. S. (1987b). An algorithm for cluster identification. *IEEE Trans. Syst., Man Cybernet.* **SMC-17**(4), 696–699.

Kusiak, A., Vannelli, A., and Kumar, K. R. (1986). Clustering analysis: Models and algorithms. *Control Cybernet* **15**(2), 139–154.

Lee, R. C. T. (1981). Clustering analysis and its applications. In "Advances in Information Systems Science" (J. T. Tou, ed.). Plenum, New York.

McAuley, J. (1972). Machine grouping for efficient production. *Prod. Eng.* February, pp. 53–57.

McCormick, W. T., Schweitzer, P. J., and White, T. W. (1972). Problem decomposition and data reorganization by clustering technique. *Oper. Res.* **20,** 992–1009.

Seifoddini, H. (1986). Improper machine assignment in machine-component grouping in group technology. *Proc. Fall Ind. Eng. Conf.,* Boston, *1986,* pp. 406–409.

Seifoddini, H., and Wolfe, P. M. (1986). "Application of the similarity coefficient method in group technology." *IIE Trans.* **18**(3), 271–277.

Siljak, D. D., and Sezer, M. F. (1984). Nested decomposition into weakly coupled components. *IFAC World Congr., 9th,* Budapest, Hungary, *1984.*

Slagle, J. L., Chang, C. L., and Heller, S. R. (1975). A clustering and data reorganization algorithm. *IEEE Trans. Syst., Man Cybernet.* **SMC-5,** 125–128.

Tou, J. T., and Gonzalez, R. C. (1974). "Pattern Recognition Principles." Addison-Wesley, Reading, Massachusetts.

CHAPTER 7

A Practical Method to Represent Uncertainty in Flexible Manufacturing Systems Justification

Alan M. Behrens
Northern Natural Gas Company
Omaha, Nebraska

and

F. Fred Choobineh
University of Nebraska at Lincoln
Lincoln, Nebraska

I. Introduction

In many areas, U.S. manufacturing has lost its edge in efficiency and quality. While many factors can be identified as possible causes of this problem, one prime candidate is the reluctance of U.S. industries to adopt new manufacturing technologies such as flexible manufacturing systems (FMSs) that would facilitate the translation of leading-edge research into competitive, quality products. While it is easy to qualitatively list the competitive advantages of FMS technology, it has often proven quite difficult for U. S. industries to economically justify their investment in these systems. One of the primary reasons cited for this difficulty has been the myopic, "short-term" view taken by many financial decision makers in their recommendations for capital investment. Traditional capital budgeting techniques often have a bias toward a short planning horizon, opting for quick returns on investment with little regard for the longer term. Thus, because of the nature of FMS investments, the traditional types of financial analyses often show them to be economic "losers."

In light of the short-term bias of traditional techniques, many authors (e.g., Blank, 1985; Kaplan, 1984) have suggested alternative approaches for the analysis of capital investments in new manufacturing technologies. The basis of these approaches is that an investment in FMS technology is not a typical investment in an improved evolutionary technology, but rather, a revolutionary, strategic investment that must look to longer-term implications in its justification. Thus, it has been suggested that analysts developing a justification for FMS must consider not only the tactical, short-term factors such as the present worth of identifiable costs and net benefits, but, also, the longer-term strategic factors such as competitive position, high-technology expertise, market leadership, and so on. Because of their qualitative nature, these factors resist the application of traditional engineering economic methods, and require more innovative approaches to their analysis. Writers such as Choobineh (1986), Padmanabhan (1989), and others have suggested a methodology where a traditional present value analysis is employed for the tactical factor analysis, and strategic factors are analyzed through a scoring and weighting scheme, the results of the tactical analysis becoming one of the factors in the strategic analysis. This approach eliminates the inherent problem of forcing dollar quantification of strategic factors far into the future.

One critical area that has been overlooked in these approaches, however, is the consideration of the inherent uncertainty present when conducting justification studies. Intuitively, the justification of a revolutionary technology such as FMS by an organization will be accompanied by a great deal of uncertainty, obviously more than in economic analyses of mature technologies. The increased uncertainty surrounding FMS studies can be attributed to several factors present in the environment of revolutionary change including

- Unproven, new technologies
- Revised ways of conducting business
- Longer planning horizons
- Greater competition
- Revamped internal procedures
- Changes in the level and use of personnel and labor
- Unknown customer responses
- New training requirements
- New interfaces to existing manufacturing systems
- More difficult economic estimates

In addition to the uncertainty of required parameter estimates, there is the added general uncertainty inherent in the consideration of revolutionary revisions to the way a company conducts its business. If studies are conducted without the explicit consideration of uncertainty, the decision maker relying on the analyses will have no idea of how "solid" are the analysis outputs that form the basis of the decision. Also, the decision maker should not rely solely on the value of a performance measure, but also on the uncertainty inherent in the measure itself. The method by which uncertainty is captured may affect the outcome of the study by changing the value of performance measures. If uncertainty is not considered, the decision maker will be denied a key piece of information.

An obvious problem in the development of an analysis which explicitly recognizes uncertainty is the choice of a representation or paradigm which provides an adequate level of rigor without unduly restricting or complicating the analysis procedure. Given the typical organizational environment in which an FMS justification study is conducted, a number of general requirements for a practical uncertainty approach can be stated:

- It must be easy to learn, apply, and explain.
- It should be adequately rigorous, but not overly intensive computationally.
- It should allow for easy visualization and simple graphical representation.
- Required estimates should be straightforward and uncomplicated.
- It should provide for the combination of multiple estimates.
- It should also include an adequate ranking methodology for comparison of results.
- It should be implementable on computers in a straightforward fashion.

While a number of different paradigms have been used to represent uncertainty (see Schum, 1986), one of the more versatile representations that addresses many of the requirements are possibility distributions.

II. A Brief Background on Possibility Distributions

Possibility distributions are a more flexible and less restrictive alternative than the traditional probability approach in the representation of uncertainty. Possibility distributions have their theoretical basis in the idea of intervals rather than relative frequency. Intervals are the simplest representation of uncertainty. Specifying an interval for a parameter or measure indicates that the parameter or measure can take on values only within that interval. Figure 1 shows an interval representation. As can be seen, an interval is specified by a lower bound *a* and an upper bound *b*. The values within the interval are completely "possible," and

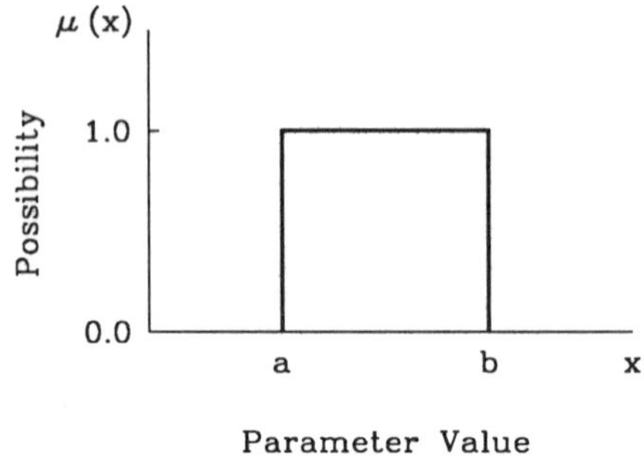

Figure 1. Interval representation.

thus have a possibility of 1, as shown on the vertical axis. Values outside the interval are "impossible" and thus have a possibility of 0. The function which maps parameter values to a level of possibility is called the characteristic function, $\mu(x)$.

Note that the characteristic function specifies the possibility of the parameter value being in the specified interval. It does not indicate how probable it is that the value is in the interval, nor does it give any indication of which particular value in the possible interval is the most likely to occur. The width of the interval, however, gives an indication of the relative uncertainty of the parameter in question. The narrower the interval, the less uncertainty that exists. Intervals, themselves, are quite useful for modeling uncertainty in minimal information situations (Choobineh and Behrens, 1990). The use of interval values in economic calculations requires the use of interval arithmetic. The basic operations of interval arithmetic are

$$\tilde{X}_1 + \tilde{X}_2 = [X_{1L} + X_{2L}, X_{1U} + X_{2U}]$$

$$\tilde{X}_1 - \tilde{X}_2 = [X_{1L} - X_{2U}, X_{1U} - X_{2L}]$$

$$\tilde{X}_1 * \tilde{X}_2 = [\text{miin}(X_{1L} * X_{2L},\ X_{1L} * X_{2U}, X_{1U} * X_{2L},\ X_{1U} * X_{2U}),$$
$$\max(X_{1L} * X_{2L},\ X_{1L} * X_{2U}, X_{1U} * X_{2L},\ X_{1U} * X_{2U})]$$

$$\tilde{X}_1 / \tilde{X}_2 = [X_{1L}, X_{1U}] * [1/X_{2U}, 1/X_{2L}] \text{ if } 0 \notin X_2$$

where

$$\tilde{X} = [X_L, X_U] = [x \mid X_L \leq x \leq X_U]$$

One problem with using interval arithmetic in calculations is that the distributive property does not hold for interval arithmetic operations. As a result, a procedure called the vertex method (Dong and Shah, 1987) must be used when performing calculations. The vertex method is described in more detail in the appendix.

While intervals can be very useful, they are also an unrealistic model for many situations. The basic premise of an interval is that the possibility of a parameter in question can instantly change from being totally possible to totally impossible at the lower and upper boundary values. Often, a more intuitively appealing assumption is that the transition from impossible to possible is gradual, and follows a particular functional form. This is precisely the idea of possibility distributions.

Possibility distributions are an extension of interval analysis that allows intervals with possibilities between 0 and 1, inclusive. Possibility, itself, is really a measure of the subjective "feel" or heuristic "hunch" of how possible it is that a particular value lies within a certain set or range of values. Possibility distributions are also known as fuzzy sets and were originated by Zadeh (1965, 1978).

Figure 2 shows a possibility distribution for a particular parameter. Note that the original interval $[a, b]$ is still present at a possibility of 1. However, the transition from possible to impossible is no longer instantaneous, but follows a linear function. Often this function is not entirely specified, but is derived based on intervals estimated at possibility levels other than 1.0. Note in Figure 2, that intervals are also specified for possibility levels of 0.5 and 0.0, and the characteristic function is derived by connecting the interval end points. This particular possibility distribution is an example of a trapezoidal possibility distribution. The nested intervals at specified possibility levels are called *alpha cuts*. Alpha cuts are defined by the following:

$$\tilde{I}_\alpha = [i \in R^+ \mid \mu_{\tilde{I}}(i) \geq \alpha,\ 0 \leq \alpha \leq 1]$$

where

R^+ = the set of positive real numbers

$\mu_{\tilde{I}}$ = the degree of belongingness, i, of the set $\tilde{I}$

A special class of possibility distributions that is suitable for justification applications has two basic axiomatic requirements. First, the transition intervals must be nested so that each interval with a higher possibility is contained in an interval with a lesser possibility. Additionally, each possibility distribution must have a parameter interval with a convex and maximum possibility of 1.0. These requirements make possibility distributions a normal measure, and give a consistent basis of comparison between distributions. The width and skewness of a particular distribution give an intuitive visual and graphical indication of the uncertainty an expert feels is present in a given parameter estimate.

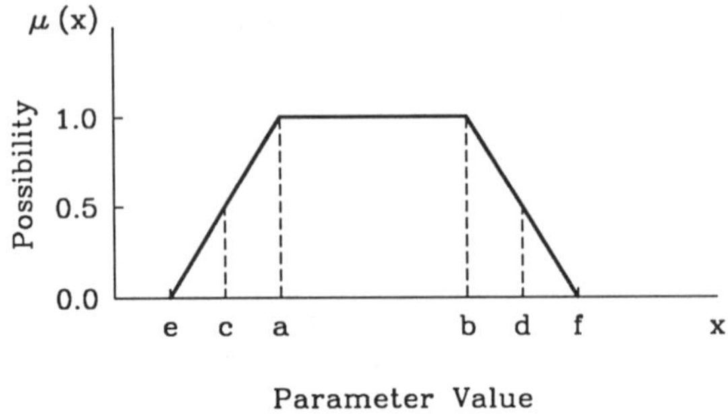

Figure 2. Possibility distribution.

Computations involving possibility distributions require the use of interval arithmetic on the interval values at each specified alpha cut defining the particular possibility distribution. The resultant possibility distribution is formed out of the interval-valued results of the computation at each alpha level. Further details of the computational aspects of applying possibility distributions are discussed in Choobineh and Behrens (1990).

Possibility distributions have a number of desirable characteristics which make them ideal for application in highly uncertain and relatively sparse information situations. These characteristics include:

1. Estimation and interpretation of possibility distributions is straightforward and requires a relatively short learning period. The ease of use of possibility distributions can be easily enhanced by attaching a linguistic interpretation to the various alpha intervals defining a possibility distribution. For example, it is reasonable to refer to the interval with a possibility of 1.0 as the *most likely* interval, the interval with a possibility of 0.0 as the *widest possible* interval, and the interval at a possibility level of 0.5 as the *mid-likely* interval. Employing linguistic terms, in practice, gives an intuitive meaning to the various possibility levels specifying a distribution and makes the estimation and interpretation of the distribution much easier. In most circumstances, three possibility intervals are more than adequate to specify a distribution.

2. Graphic representations of possibility distributions are easy to construct and display, given the specified intervals. Interpretation and comparison of possibility distributions is greatly enhanced through graphic representation.

3. Multiple possibility distribution estimates of a given parameter can be combined to form a single distribution that captures the combined essence of each estimator's input. Two combination methodologies for possibility distributions are discussed in Shafer (1984). One is called *Zadeh's Rule of Combination*

and is described by Equation (1), while the other is known as the *Likelihood Rule of Combination* and is given by Equation (2).

$$f(s) = \min(f_1(s), f_2(s))/\sup \min(\{f_1(s')f_2(s') \mid s' \text{ in } S\}) \quad (1)$$

$$f(s) = (f_1(s)f_2(s))/\sup(\{f_1(s')f_2(s') \mid s' \text{ in } S\}) \quad (2)$$

where

$$S = \text{universe of possible values}$$

$$f(s) = \text{possibility of } s$$

An alternative approach that simplifies the computer implementation of distribution combination and also allows a relative weighting of each respective expert's input is given in Behrens (1989). This approach is given by Equation (3).

$$\tilde{I}_\alpha = [\,(\sum_i X_{Li}\,\beta_i)/\sum_i \beta_i,\ \sum_i (X_{Ui}\,\beta_{iV})/\sum_i \beta_i] \quad (3)$$

where

$\tilde{I}_\alpha$ = alpha cut of combined possibility distribution

X_{Li} = lower interval value of *i*th possibility distribution

X_{Ui} = upper interval value of *i*th possibility distribution

β_i = weighting of *i*th possibility distribution

4. The computational aspects of possibility distributions lend themselves to fairly easy implementation on computers. The spreadsheet/graphics type of microcomputer software is particularly useful for prototyping and implementing models involving possibility distributions because of the combined computational and graphical facilities available within this type of software.

5. An important aspect of any uncertainty representation is the basis by which the results of one particular case or analysis can be compared or ranked with the results of another alternative. There are several ranking methodologies defined within the possibility distribution framework. Many of these approaches are discussed in Behrens (1989) and Chen (1985). One of the most intuitive and straightforward approaches is the centroidal method, where the ranking value of a given possibility distribution is determined by the parameter value located at the x coordinate of the centroid.

The following section illustrates the practical application of the possibility distribution representation in an FMS justification study.

III. Application of Possibility Distributions in Flexible Manufacturing Systems Justification

The characteristics of possibility distributions make them an ideal paradigm with which to incorporate uncertainty into both the tactical and strategic phases of an FMS justification. The basic FMS justification approach used here will be as described in Choobineh (1986), where the tactical justification portion consists of a present-value analysis of shorter term, quantifiable economic factors. These factors can include first costs:

- Initial investment
- Installation costs
- Engineering costs
- Tooling, fixtures, etc.
- Facility modifications

and annual benefits:

- Direct labor savings
- Material savings
- Energy savings
- Standard overhead reduction
- Inventory cost reduction
- Quality improvements
- Other annual benefits/costs

Additional factors that must be obtained include interest rate estimates, tax rates, and equipment life.

The steps for conducting the tactical portion of the possibilistic justification study are as follows:

1. For each of the economic factors to be incorporated into the tactical portion of the analysis, a possibility distribution must be estimated. Alpha cuts at the 0 and 1.0 possibility levels are the minimum required to form a distribution. To make the estimation process easier, it is usually desirable to attach a linguistic interpretation to the various possibility levels desired, such as *widest possible interval* for the 0.0 possibility level, *most likely interval* for the 1.0 possibility level, and *mid-likely interval* for the 0.5 possibility level. The intervals, thus defined, can be formed into possibility distributions for visual comparison and interpretation. If multiple estimates of a factor are obtained, they must be combined into a single distribution using one of the combination techniques for possibility distributions.

2. At each alpha cut, the desired net present value (*NPV*) should be determined using interval arithmetic operations to perform the necessary calculations. Calculating the present value using interval arithmetic will require the use of the vertex method, as outlined in the appendix. The results, as calculated at each possibility level, can be combined to form a possibility distribution of the NPV for each alternative.

3. Given the *NPV* possibility distributions of the tactical factors, a scoring possibility distribution must be derived for each alternative so that the results of the tactical analysis can be factored into the strategic analysis. The *NPV* scoring distribution is derived by mapping the *NPV* values of each alpha cut onto the range of possible scoring values using the mapping function given in Equation (4). The range must be the same as used in the strategic analysis.

$$X^s = 1 + (S_{\max} - S_{\min})\,[(X^p - X_{\min})/\,(X_{\max} - X_{\min})] \quad (4)$$

where

X^s = resultant *NPV* scoring value

X^p = given *NPV* value

$X_{\min}$ = minimum *NPV* value across all alternatives

$X_{\max}$ = maximum *NPV* value across all alternatives

$S_{\min}$ = minimum scoring value

$S_{\max}$ = maximum scoring value.

The tactical scoring distribution, thus derived, is then incorporated into the strategic analysis.

The strategic portion of a possibilistic FMS justification analysis uses a weighting and scoring methodology to rank various longer-term, qualitative factors and the present value results of the tactical analysis for each alternative. Many of these qualitative factors are very difficult or impossible to analyze using standard economic analysis approaches because they resist easy quantification. Ignoring these factors, however, is essentially the same as assigning a zero value to them, thus making any analysis "precisely wrong" rather than "vaguely right" (Kaplan, 1986), and possibly leading to the unwarranted rejection of a leading-edge FMS alternative. In addition to the NPV results of the tactical portion of the analysis, examples of strategic factors that should be accounted for include

- Competitive advantage of each alternative
- Technological advantages of each alternative
- Industry leadership
- Product development time

In the strategic analysis of Choobineh (1986), each relevant factor is given a weight that indicates the relative significance that the factor has to the organization. Alternative FMS proposals are then scored on each of the factors. The relative weight of each factor is applied to its score, and the results are summed for each proposed alternative. In this way the strategic implications of each alternative are ranked.

One aspect of strategic analysis is the determination of the weight that a particular organization should assign to each strategic factor. This weighting can be very uncertain, since it will depend on many qualitative internal and external aspects of the environment. Also, within the organization, there will be many differences of opinion as to the relative weight that each factor should be assigned. Combining these opinions in a consistent manner to form a consensus weighting is also a problem that must be addressed.

Another aspect of the strategic evaluation process is the determination of the grading or scoring value to be assigned to each factor for a particular alternative. It will be quite difficult to assign a single scoring number for a certain factor of a given alternative with any certainty. In addition, it is often desirable to solicit several independent scoring estimates. This will lead, as above, to the problem of melding multiple inputs for scoring into a single, consensus value.

Possibility distributions can effectively represent the uncertainties in the strategic evaluation process. Rather than giving single-valued estimates for the weighting and scoring of each relevant strategic factor, experts can be asked to give the *most likely, mid-likely,* and *widest possible* range of values, corresponding to the 0.0, 0.5, and 1.0 levels of possibility. The estimates can then be used in the possibility distribution framework.

The procedure for conducting a possibilistic strategic analysis of alternative FMS proposals is as follows:

1. Develop a list of strategic factors important to the longer-term viability and success of the organization. Factors both internal and external to the organization should be considered. The final list is probably a function of top management input.

2. Solicit from selected internal and/or external experts the possibility distribution estimates of the relative weights that should be assigned to each of the strategic factors, including the results of the tactical portion of the analysis. Weight estimates should be solicited for possibility levels of 1.0 and 0.0 at a minimum. For those weighting factors where multiple experts have been solicited, combine the respective possibility estimates into a single distribution using an appropriate combination technique.

3. For each proposed FMS alternative, solicit a scoring possibility distribution for each strategic factor from selected experts. Combine multiple estimates

of possibility distributions into a single distribution, as required, using an appropriate combination technique.

4. Multiply each scoring distribution for each factor by its respective weighting distribution using interval multiplication at each alpha level. This operation will form a weighted scoring distribution for each factor for each alternative. The respective factor distributions for each alternative can then be summed at each alpha level using interger addition to form a final ranking possibility distribution for each alternative. The final ranking distribution computation is given by the following equation at each alpha cut:

$$\tilde{I}_i^\alpha = \sum_j \tilde{w}_j \, \tilde{s}_{ij} \tag{5}$$

where

$\tilde{I}_i^\alpha$ = ranking interval value at possibility value α for alternative i

$\tilde{w}_j$ = weight interval of jth factor at possibility α

$\tilde{s}_{ij}$ = scoring interval of jth factor for alternative i

These intervals form the final ranking distribution for each alternative. The ranking value of each alternative can then be determined by using one of the ranking methodologies for possibility distributions such as the centroidal approach. This approach computes a ranking value by finding the weighted scoring value at the centroid of each final ranking distribution. Graphical displays of the intermediate steps in this process give a visual feel of the relative uncertainty in the estimates and computations and can be used to compare factors and weights for each alternative.

IV. Numerical Example

Company XYZ wishes to evaluate two alternative FMSs for possible incorporation into its operation. To aid in the selection process, the company has decided to conduct a possibilistic justification study of the two alternatives. To analyze the tactical aspects of these alternatives, XYZ has solicited estimates of possibility distributions of various relevant parameters from specialists within the company. Table 1 gives the estimates in terms of possibility intervals at each specified alpha level.

XYZ is also quite uncertain about the interest rate to use on this project. Therefore, it has retained two outside consultants to estimate possibility

TABLE 1
Tactical Analysis Possibility Estimates

		First cost		Annual benefits	
	Possibility	Atlernative 1	Alternative 2	Alternative 1	Alternative 2
Most likely value(s)	1.0	(500, 700)	(400, 500)	(500, 700)	(400, 500)
Mid-likely range	0.5	(400, 800)	(300, 600)	(400, 800)	(300, 600)
Widest possible range	0.0	(300, 900)	(200, 700)	(300, 900)	(200, 700)

distributions of possible interest rates to use in the analysis. XYZ, however, has somewhat more confidence in expert 1 than in expert 2 and wishes to weigh expert 1's estimates twice as heavily into the analysis. Table 2 gives the estimates of interest rates furnished by the outside experts.

In addition to these possibilistic estimates, other important parameters for the tactical analysis are

$$\text{Project life} = 10 \text{ yr}$$

$$\text{Tax rate} = 50\%$$

$$\text{Depreciation} = \text{Straight line}$$

The first step in the possibilistic analysis is the combination of possibility distributions for those parameters where there are multiple estimates. In this case, two independent estimates of applicable interest rates have been given, as shown graphically in Figures 3 and 4. Note the differing perceptions of the experts as embodied in the respective distributions. Where expert 1 has estimated a symmetric distribution of possibilities, expert 2's estimate is skewed toward the left. To combine these, the weighted interval approach as outlined in Behrens (1989) is used. The resultant final interest possibility distribution shown in Figure 5 captures the essence of each estimator's perception of uncertainty.

TABLE 2
Interest Rate Estimates

	Possibility	Expert 1 (wt. = 2)	Expert 2 (wt. = 1)
Most likely value(s)	1.0	(10, 10)	(11, 12)
Mid-likely range	0.5	(8, 12)	(9, 13)
Widest possible range	0.0	(6, 14)	(7, 14)

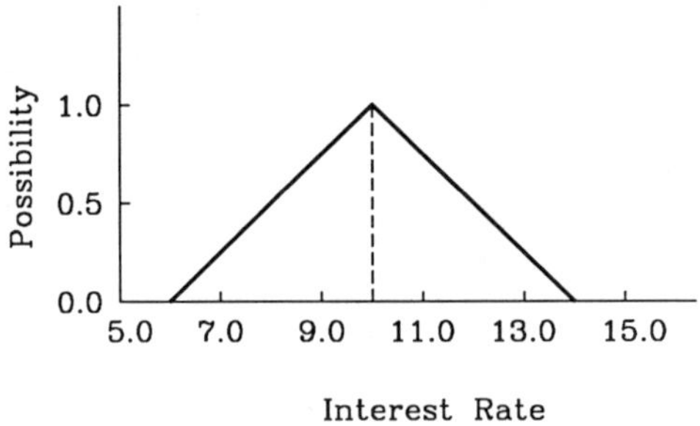

Figure 3. Expert 1 interest rate.

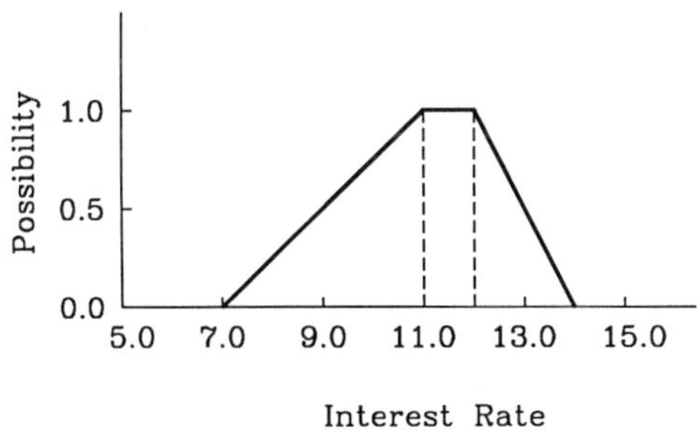

Figure 4. Expert 2 interest rate.

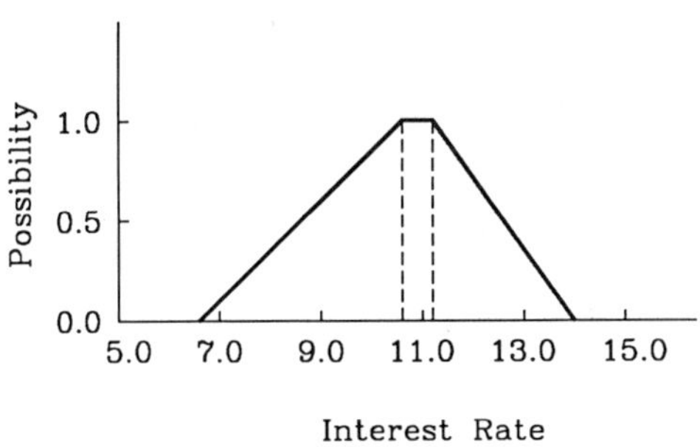

Figure 5. Combined interest rate.

TABLE 3
After Tax Benefit Flow

	Alternative 1					Alternative 2				
Possibility	BTBF	Depreciation	Taxable income	Income taxes	ATBF	BTBF	Depreciation	Taxable income	Income taxes	ATBF
0.00	100.00	30	10.00	5.00	95.00	50.00	20	−20.00	−10.00	60.00
0.50	200.00	40	120.00	60.00	140.00	75.00	30	15.00	7.50	67.50
1.00	300.00	50	230.00	115.00	185.00	100.00	40	50.00	25.00	75.00
1.00	350.00	70	300.00	150.00	200.00	150.00	50	110.00	55.00	95.00
0.50	400.00	80	360.00	180.00	220.00	175.00	60	145.00	72.50	102.50
0.00	450.00	90	420.00	210.00	240.00	200.00	70	180.00	90.00	110.00

TABLE 4
Net Present Value Distribution

	Alternative 1						Alternative 2			
Possibility	Interest rate	P/A factor	PW ATBF	First cost	*NPV*	Scoring distribution	PW ATBF	First cost	*NPV*	Scoring distribution
0.00	6.60	7.155	495.53	−900	−404.47	1	312.97	−700	−387.03	10
0.50	8.60	6.532	771.98	−800	−28.02	187	372.20	−600	−227.80	88
1.00	10.60	5.989	1080.43	−700	380.43	388	438.01	−500	−61.99	170
1.00	11.20	5.840	1197.87	−500	697.87	545	568.99	−400	168.99	284
0.50	12.60	5.514	1437.09	−400	1037.09	712	669.55	−300	369.55	383
0.00	14.00	5.126	1717.27	−300	1417.27	900	787.08	−200	−587.08	490

TABLE 5
Planner Responses

	Expert 1 weighting distributions (1-100)					Expert 2 weighting distribution				
Possiblity level	Competitive advantage	Technology experience	Industry leadership	Development time	*NPV*	Competitive advantage	Technology experience	Industry leadership	Development time	*NPV*
0.00	30	50	10	70	30	70	40	10	50	40
0.50	45	60	20	80	40	75	45	15	65	45
1.00	55	70	30	90	50	80	50	20	80	55
1.00	57	80	40	90	55	80	60	20	80	57
0.50	58	85	45	94	58	85	65	30	85	58
0.00	60	90	50	98	60	90	70	40	90	60

Given the preceding possibility distribution estimates, the net present value (*NPV*) possibility distribution is determined by calculating the interval-valued *NPV* at each alpha level. This is given by the following:

$$NPV = FC + ATBF(P/A,N,i)$$

where

$$FC = \text{interval-valued first cost}$$

$$(P/A,N,i), = \text{interval-valued present value calculated over period } N, \text{ at interest } i$$

$$ATBF = \text{annual after-tax benefit flow interval}$$

ATBF is defined as follows:

$$ATBF = BTBF \text{ (before tax benefit flow)} - IT \text{ (income taxes)}$$

where

$$IT = (BTBF - \text{depreciation}) \times \text{Tax Rate}$$

$$\text{Depreciation} = \text{First cost/Life of investment}$$

Table 3 shows the resulting calculation of *ATBF* at each alpha level based on the parameters estimated above.

The present values of these annual benefit flows are then calculated over the life of the investment. The interval-valued calculations require the use of the vertex method, as outlined in the appendix. The resulting annual benefits are added to the first cost of each alternative to give an *NPV* possibility distribution for each alternative. The resulting *NPV* values are shown in Table 4 for each alternative at the same respective possibility interval levels as in the previous tables.

TABLE 6
Combined Weighting Distributions (1-100)

Possibility interval		Competitive advantage	Technology experience	Industry leadership	Development time	*NPV*
Widest possible	low	58.0	43.0	10.0	56.0	37.0
Mid-possible	low	66.0	49.5	16.5	69.5	43.5
Most possible	low	72.5	56.0	23.0	83.0	53.5
Most possible	high	73.1	66.0	26.0	83.0	56.4
Mid-possible	high	76.9	71.0	34.5	87.7	58.0
Widest possible	high	81.0	76.0	43.0	92.4	60.0

TABLE 7
Combined Scoring Distribution

	Alternative 1					Alternative 2				
Possibility level	Competitive advantage	Technology experience	Industry leadership	Development time	*NPV*	Competitive advantage	Technology experience	Industry leadership	Devlopment time	*NPV*
0.00	275	300	600	550	1	300	350	400	400	10
0.50	338	350	650	675	187	375	450	450	450	88
1.00	400	400	700	800	388	450	550	500	500	170
1.00	425	500	750	800	545	500	700	550	650	284
0.50	475	575	788	850	712	538	775	600	725	383
0.00	525	650	825	900	900	575	850	650	800	490

TABLE 8
Weighted Scoring Distribution

	Alternative 1					Alternative 2				
Possibility level	Competitive advantage	Technology experience	Industry leadership	Development time	*NPV*	Competitive advantage	Technology experience	Industry leadership	Development time	*NPV*
0.00	15950	12900	6000	30800	37	17400	15050	4000	22400	355
0.50	22275	17325	10725	46913	8124	24750	22275	7425	31275	3836
1.00	29000	22400	16100	66400	20776	32625	30800	11500	41500	9095
1.00	31067	33000	19500	66400	30737	36550	46200	14300	53950	16017
0.50	36528	40825	27169	74545	41318	41334	55025	20700	63583	22212
0.00	42525	49400	35475	83160	54000	46575	64600	27950	73920	29419

The present value of the *ATBF* is calculated for both the high and low interest values on each possibility interval. This calculation is an application of the vertex method, as mentioned above.

Using the *NPV* distributions, a scoring distribution is then calculated for each alternative by mapping the respective *NPV* distributions onto the range of scoring values (1–900) to be used in the strategic analysis. This mapping is completed using Equation 4 and shown in Table 4.

For the strategic portion of the study, company XYZ management has drawn up the following list of strategic factors deemed important to the long-term survival of the company. These factors are

- Competitive advantages
- Experience with alternative or new technology
- Industry leadership
- Product development time
- *NPV* of tactical analysis

Management then asked two of its best long-range planners to provide a possibility distribution of the relative weight (1–100) that should be assigned to each factor.

Table 5 shows the responses of each planner. Management has decided that planner 1's inputs should carry more weight (0.7) than planner 2's estimates (0.3). Table 6 shows the combined weighting distribution after combination by the method of Equation (3).

The two alternatives were then scored possibilistically over the range of scoring values by selected experts for each factor. Where appropriate, the estimates were combined, giving the final scoring distributions as shown in Table 7. The *NPV* score came from the tactical portion of the analysis completed above.

TABLE 9
Final Ranking Distributions

Possibility interval		Alternative 1	Alternative 2
Widest possible	low	65687	59205
Mid-possible	low	105362	89561
Most possible	low	154676	125520
Most possible	high	180705	167017
Mid possible	high	220385	202853
Widest possible	high	264560	242464
Centroidal Ranking Value		166079	149032

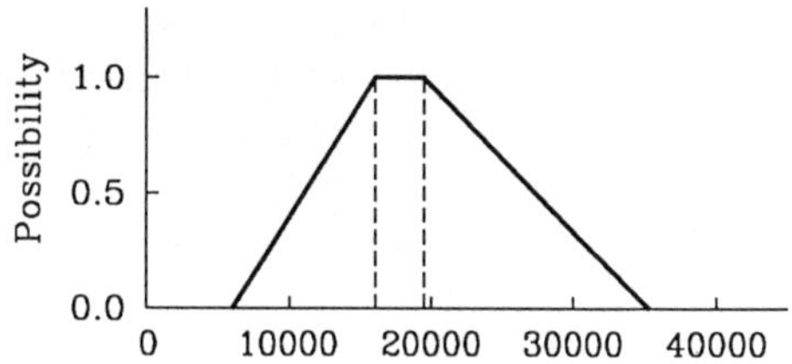

Figure 6. Industry leadership, alternative 1.

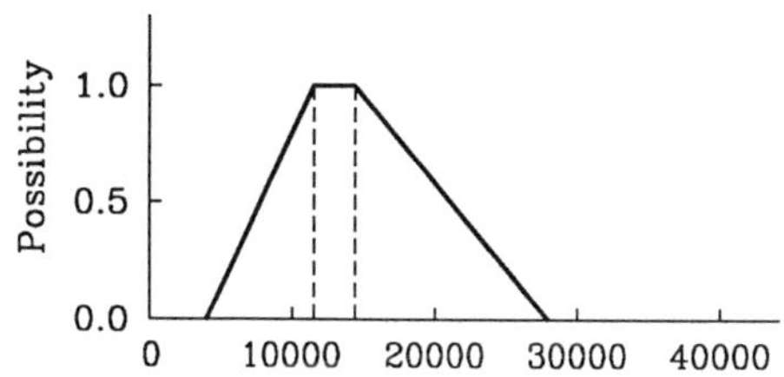

Figure 7. Industry leadership, alternative 2.

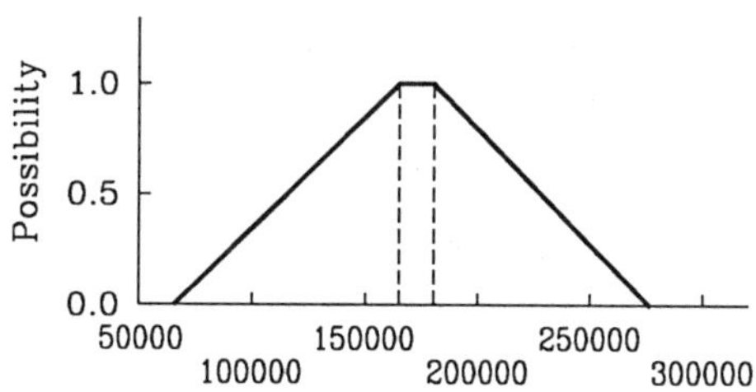

Figure 8. Alternative 1, final combined.

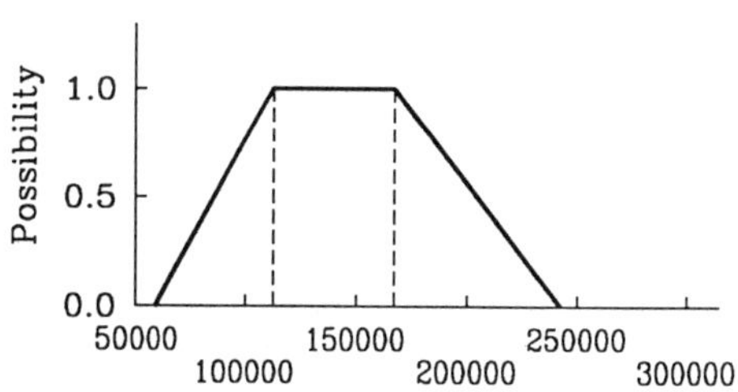

Figure 9. Alternative 2, final combined.

Using Equation (5), the weighted scoring distribution is calculated for each factor for each alternative. The results are given in Table 8. Table 9 shows the final ranking distributions along with the final centroidal ranking value for each alternative. Figures 6 and 7 show graphically the weighted scoring possibility distributions for the industry leadership factor. All factors can be displayed in this manner to graphically compare the magnitude and uncertainty of each strategic factor by alternative.

The final centroidal ranking value shows alternative 1 to be the choice of FMS alternatives. The final centroidal value, in effect, captures the results of all of the analyses and the uncertainty in one number for comparison. The final ranking distributions, shown in Figures 8 and 9, are very similar for both alternatives as far as shape and spread, indicating very similar overall uncertainty characteristics between the alternatives, although alternative 1 has a somewhat narrower range at the 1.0 possibility level. Alternative ranking methodologies such as discussed in Behrens (1989) or Chen (1985) can yield somewhat different final ranking values.

V. Conclusion

This paper has demonstrated a methodology for incorporation of uncertainty into both the tactical and strategic analysis of FMS justification. Uncertainty in parameter estimation is a very important factor to consider in any analysis, but it is particularly important when the analysis is dealing with the introduction of a revolutionary technology over a long planning horizon such as FMS. Failure to explicitly consider uncertainty ignores what should be a major consideration in the analysis of any situation.

The uncertainty paradigm used in this paper offers several advantages over more traditional approaches. Specifically, possibility distributions capture the essence of uncertainty as expressed by the estimators in an intuitive and visual manner. This approach also provides a mechanism for combining expert options in a consistent and logical fashion to form a consensus estimate that captures the combined uncertainty of the input estimates. In addition, the low information requirements of this representation allow the analysts and decision makers to concentrate on the problem at hand, not on the nuances of the particular methodology being employed. Finally, this particular approach to uncertainty lends itself very well to prototyping on microcomputers using spreadsheet/graphics types of software, and it is also straightforward to translate the uncertainty model into a procedural type of language for a more formal system.

Appendix: The Vertex Method

The major computational difficulty with interval operations is that distributivity does not hold for interval operations. In practical terms, this means that the order of calculation of a function involving interval numbers determines the functional results. Dong and Shah (1987) suggested a procedure called the vertex method that allows distributivity to hold in interval arithmetic.

A function of interval variables is an interval variable itself, and it is defined as

$$\begin{aligned}\tilde{Y} &= f(X_1, X_2, \ldots, X_n) \\ &= \{f(x_1, x_2, \ldots, x_n) \mid x_1 \in X_1, x_2 \in X_2, \ldots, x_n \in X_n\}\end{aligned}$$

The n interval variables form an n-dimensional rectangular region $X_1, X_2, \ldots X_n$ with $N = 2^n$ vertices. The coordinates of vertices are combinations of n pairs of the limits of the interval variables. Let v_j, $j = 1, \ldots, N$ be the jth vertex of an n-dimensional rectangle. Assuming the function f is continuous in the n-dimensional region, and there are no extreme points in this region, the interval variables $\tilde{Y}$ can be obtained by

$$\tilde{Y} = [\min_j (fv_j), \max_j (fv_j)] \; j = 1, \ldots, N$$

In practical terms, the vertex method means that, if the function is continuous, all combinations of variables must be considered in the function, and the minimum and maximum functional values form the interval output.

References

Behrens, A. M. (1989). An approximate reasoning model for natural gas pricing. Unpublished doctoral dissertation, University of Nebraska, Lincoln.

Blank, L. (1985). The changing scene of economic analysis for the evaluation of manufacturing system design and operation. *Eng. Econ.* **30** (3) 227–244.

Chen, S. (1985). Ranking fuzzy numbers with maximizing set and minimizing set. *Fuzzy Sets Sys.* **17,** 113–129.

Choobineh, F. (1986). Economic justification of flexible manufacturing systems. *Proc. ASME Int. Comput. Eng. Conf. Exhib.* pp. 269–279.

Choobineh, F., and Behrens, A.M. (1990). "Use of Intervals and Possibility Distributions in Economic Analysis," Working Paper. Department of Industrial Engineering, University of Nebraska, Lincoln.

Dong, W., and Shah, H. (1987). Vertex method for computing functions of fuzzy variables. *Fuzzy Sets Syst.* **24,** 65–78.

Kaplan, R.S., (1984). Yesterday's accounting undermines production. *Harv. Bus. Rev.* July-August, pp. 95–101.

Kaplan, R. S. (1986). Must CIM be justified by faith alone? *Harv. Bus. Rev.* March-April, pp. 87–95.

Padmanabhan, S., (1989) A tandem expert support system as justification for a flexible manufacturing system. *J. Manuf. Syst.* **8** (3) 195–205.

Schum, D. A. (1986). Probability and the process of discovery, proof, and choice. *Boston Univ. Law Rev.* **66,** Nos. 3 & 4.

Shafer, G. (1984). "Belief Functions and Possibility Measures," Working Paper No. 163. University of Kansas, School of Business, Lawrence.

Zadeh, L. A. (1965). Fuzzy sets. *Inf. Control* (8) 338–353.

Zadeh, L. A. (1978). Fuzzy sets as a basis for a theory of possibility. *Fuzzy Sets Syst.* **1,** 3–28.

CHAPTER 8

Design of Flexible Assembly Systems Using Simulation Modeling

Jim Browne and J. Feargal Timon
CIM Research Unit
University College
Galway, Republic of Ireland

HANDBOOK OF FLEXIBLE MANUFACTURING SYSTEMS

I. Introduction

A flexible manufacturing system (FMS) is an integrated, computer-controlled complex of automated material handling devices and computer numerically controlled (CNC) machine tools that can simultaneously process medium-sized volumes of a variety of part types (Stecke, 1983). An FMS therefore combines many different automation concepts and technologies. They include numerical control (NC) and computer numerical control (CNC), direct numerical control (DNC) over the material handling system and individual machine tools, and automated material handling.

To date there have been many successful applications of FMS. There is now increasing attention being paid to flexible assembly systems (FASs). An FAS is a computer-integrated manufacturing system consisting of automated material handling devices and workstations that can simultaneously assemble a variety of part types (Hall and Stecke, 1986). Each robotic workstation or cell in an FAS is flexible and performs a wide variety of operations enabling the system to simultaneously assemble different low-volume part types.

Indeed, it is the flexibility of the FAS workstations and transfer systems which enables it to automate batch production, by using small batch sizes and facilitating batch variety. To obtain the required degree of flexibility, great effort is needed in the design of an FAS. The fact that FASs generally call for high capital investment highlights the need for cost-effective methods for planning, design, and implementation.

There is thus a clear need for an integrated procedure for FAS design. Browne *et al.* (1985) proposed a procedure for the design of FMSs. There are three stages of activity in the design and development of an FMS, namely planning, design, and implementation (see Figure 1). At the planning stage, it

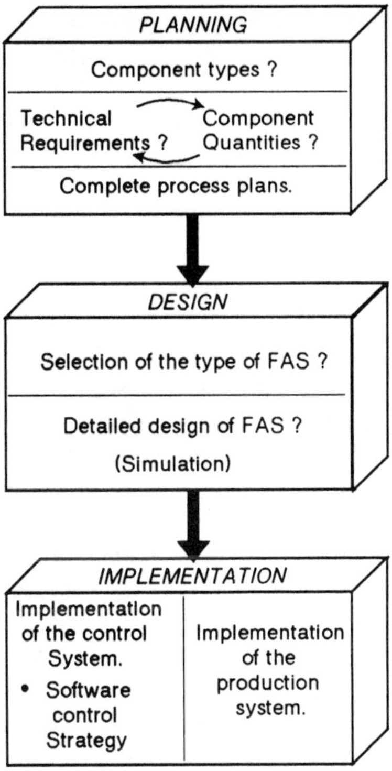

Figure 1. A three-stage design procedure for FAS.

is necessary to identify the family of products to be assembled, the quantities involved, and the technical requirements of assembly.

The second stage that is, the design stage, involves the detailed design of the FAS to meet the requirements identified in the planning stage. This design stage falls into two phases: selecting the type of FAS, in terms of the FAS architecture most likely to be suitable, and the detailed design and fine tuning of the chosen FAS.

An FAS can exhibit many different types of flexibilities. Stecke *et al.* (1984) outlined the different types of flexibility which are based on material handling, work centers, product, people, and machines. The first phase of designing an FAS involves the selection and combination of suitable flexibilities into a viable system.

Having selected the most suitable type of automated manufacturing system, the second phase involves the consideration of detail design issues relating to, for example, the number of workstations required, the type of material handling system, the location and size of work-in-progress buffers, the type of computer control, the scheduling system, and so on. Great effort is required in this phase of the design procedure in order to obtain an efficient layout, especially in the case of an FAS, where synchronization of parts is required. It is in this phase that simulation should be used to support the design of an FAS. Simulation allows the designer to experiment with different shop floor layouts, product mixes and schedules, transfer systems, and so on, without the cost that would be incurred in experimenting with a real system (Browne and Rathmill, 1989). Simulation therefore helps produce and fine tune a detailed design before the FAS is implemented.

The third stage in the proposed design procedure is the implementation phase. This divides into two separate but interrelated tasks: software design and implementation and equipment/hardware design and implementation.

In this paper we are concerned with *the use of computer simulation modeling as a tool to support the detailed design and analysis of a flexible assembly system*. Our approach is to look at an existing flexible assembly system which is in operation and see how simulation modeling can be used to support the manufacturing engineers who are in the process of changing over the line to the assembly of a new product range. The product involved is a kitchen appliance which will be assembled on a Bosch assembly line. A secondary objective of the paper is to see how well various simulation packages are able to simulate the chosen assembly system. The structure of the paper is as follows. First, we describe the structure of the assembly line and the product range which it is proposed to assemble. We then show how a simulation model of this line is developed using a number of different simulation packages, the majority of which are available in the marketplace. We conclude by offering a brief comparison of these packages and drawing some lessons on the role and usefulness of simulation modeling for assembly system analysis and design.

II. The Bosch Transfer Line for a Kitchen Appliance

A local manufacturing firm is now installing a Bosch transfer line. This will allow on-line real-time control over the assembly of one of their major products, namely a kitchen appliance. Within the plant they manufacture 95% of all their components needs. The only components bought in are internationally standard parts, such as nuts and bolts. The assembly process for the appliance consists of two assembly lines: the main assembly line on which the finished components are finally assembled, and a feeder line which produces motors. The product and the assembly process are discussed below.

A. A DESCRIPTION OF THE PRODUCT AND THE ASSEMBLY PROCESS

The product is a domestic appliance which is intended for both the European and American markets. From the point of view of the simulation model, we can represent the product range by one product although in reality there are some minor differences between products across the product range in terms of the electrical characteristics of the end product. A bill of materials showing the main components for the product is given in Figure 2. It can be seen from the diagram that the bill of materials is relatively flat. This simplifies the assembly process considerably.

The main components are partially assembled and placed on a customized pallet, before accessing the conveyor belt. All other less expensive components are stored at individual workstations. The main product consists of four components:

- The main *cabinet* or housing for the product. This is an injection-molded plastic covering and is molded within the plant.
- The *motor,* which is assembled on the feeder line.
- The *blade,* which is an extremely sharp circular piece of metal and is assembled by a robot.
- The *front cover,* which is manufactured in the injection molding area of the plant.

The main components of the motor are placed on a customized pallet before accessing the Bosch conveyor belt. The remaining components, including soldering wires, nuts, and so on, are stored at the assembly stations.

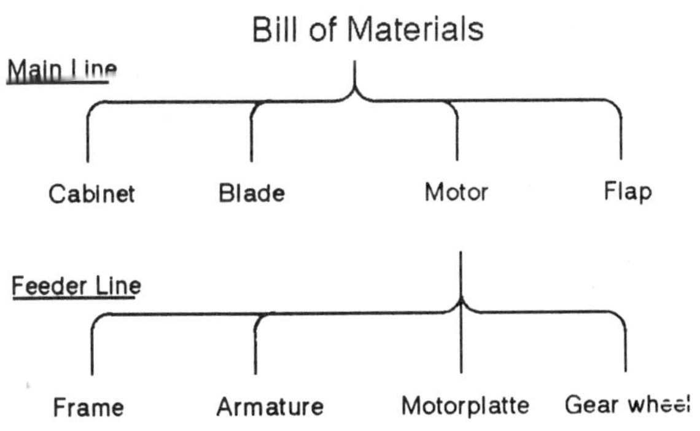

Figure 2. A bill of materials for the kitchen appliance.

1. The Feeder Line.

The feeder line is adjacent to the main assembly line. It is an asynchronous flexible assembly system with a supporting carousel. There are four main tasks involved in assembling the motor; of these the kitting and assembly operations are manual and the run-in and test stations are automatic. The different operations are as follows:

• The kitting of the main components. This involves placing the components into the specially molded groves on a customized pallet. The pallet acts as a fixture for the motor frame and holds the main components securely. The kitting operation is expected to take 18.81 s. After kitting, the pallet is placed on the conveyor where the Bosch computer control transfer system decides where it should go; either to one of the assembly stations or to an intermediate queue/carousel. All stations are illustrated in Figure 3.

• The second operation involves the total assembly of the motor. As this is a time-consuming task, there are five stations acting in parallel carrying out the operation. The five assembly workers are paid through a bonus scheme. If any of the operators is starved of work, he is paid at a rate equal to his maximum potential output. Therefore it is important that each operator is always supplied with kitted pallets. The assembly operation has been estimated to take about 126 s, but as the operators are on a bonus scheme, it is expected that this time will reduce to about 100 s.

• Having assembled the motor, all parts are checked for defects. Prior to testing, it is necessary to run-in all of the motors. This operation involves running

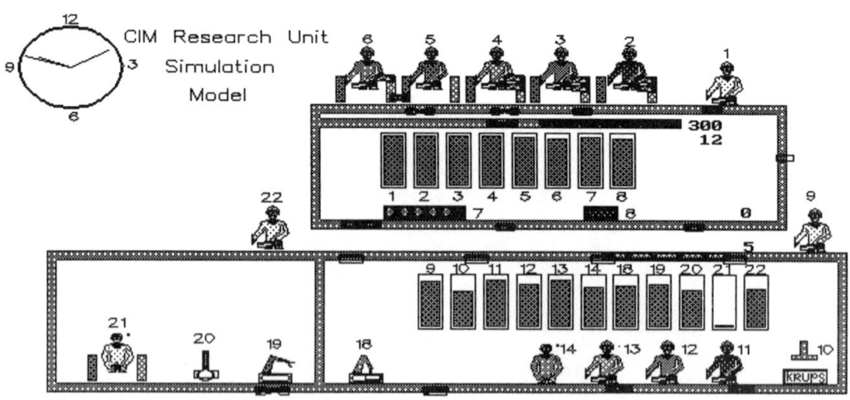

	Inter Arrival	Time in Sys	No Finished	Throughput
Motor Assembly	20.57	710.21	303	
Cabinet Assembly	22.77	426.90	258	

Figure 3. The layout of the assembly lines.

the motor for 20 s. The run-in station is automated and can process five motors simultaneously.

• The final task on the motor line is to inspect the motors for any defects. The inspection station consists of an automated sensing device and checks one part at a time. If a part is defective, it travels back around to the assembly station where it was processed for repair. Nondefective motors enter a queue where they await the kitter for the main assembly. The kitter removes the motor and places it on a pallet along with the other components required to assemble the kitchen appliance.

The feeder line is intended to feed the cabinet assembly line, in a just-in-time manner (JIT), that is, finished motors have to wait on the assembly line until a pallet on the other line is available. Then and only then is the motor pallet released for rekitting.

2. The Main Assembly Line.

The main assembly line is also an asynchronous flexible assembly system. There are 11 stations, each carrying out a unique operation. Of these 11 stations, 1 is a kitting station, 4 are hand-assembly stations, 1 is a manual repair station, 2 are robotic stations, 2 are automatic stations, and 1 is a packing station. The different operations and their associated logic are outlined below:

• The first operation on this line involves the kitting of the main parts onto a customized pallet. The pallet acts as a fixture for the cabinet and holds the other components in place. The pallet also allows access to the components from underneath, where required.

• After kitting the pallet is placed on the conveyor and travels to the emboss station. This operation is in two stages, first the embossing of the company logo onto the cabinet and, second, stamping of the date of production onto the inside of the molding. This task can be carried out before or after the manual operations.

• In the present design there are four manual assembly stations, each carrying out a different set of assembly tasks. The pallet travels to each station. If, however, the demand for finished parts changes, the number of manual stations can be altered appropriately and the assembly tasks divided out among the new set of operators.

• After the pallet has passed through the manual assembly stations, it proceeds to the two robotic cells. The first cell inserts four studs into the cabinet and then greases the parts. The second cell can carry out two operations. The robot either fits the blade to the assembly or removes it. A blade is removed when the assembly fails inspection and must be repaired.

• In all of the stations previously discussed, the only decision logic used by the conveyor system was to transfer the pallet to the next station. The logic used by the test and repair stations is a little more complex. After the blade is fixed on the assembly, the pallet is transferred to the test station. If the assembly passes inspection, it is transferred to a packing area where the finished appliance leaves the system. An assembly which fails inspection is transferred back to the blade assembly station via the crossover conveyor (see Figure 3). The blade is removed and the pallet transferred to the repair station where a repaired assembly is waiting for a pallet. The operator swops the assemblies, and the pallet with a repaired appliance is transferred back to the blade assembly station, where a new blade is attached. The pallet is then conveyed to the test station where it is treated as if it arrived there for the first time.

After a pallet is transferred to packing, the finished assembly is removed and the empty pallet is conveyed back to the kitting station. As the customized pallets are expensive, their number is limited. On the main assembly line, it is expected there will be 40 pallets and therefore a maximum of 40 parts in the system at any point in time.

III. Simulation Packages

There are many simulation packages available [see Anonymous, (1988), for a list of commercially available packages]. In this paper six simulation packages are discussed and their features highlighted in the context of the flexible assembly system under review. The packages are as follows:

- SIMAN and Cinema
- SLAM II
- MAST (SPAR and BEAM)
- SIMFACTORY
- PCModel
- FALSIM

Simulation packages use different modeling methods. There are three commonly used methods for modeling the flow of material through a manufacturing system. Two of them model discrete change systems, while the third models a continuous change system using algebraic difference/differential equations. The combination of these methods can be used to represent a combined discrete/continuous system. The three methods are as follows:

- The *process* or *network* method. The model is constructed of block commands in a linear top-down sequence, each block representing specific *process* functions such as time delays and queues.
- The *event* method. This may be used to augment or replace the block diagram component. The *event* component consists of a set of user-written sub-

routines which contain mathematical and logical expressions that define the change in the state of the system at each event time. In many simulation languages and packages these subroutines call the language's own subprograms which help effect the changes in the system.

• The two previous methods are used to model a discrete system, that is, a specific action or event causes a change in the system state at a specific time. In a *continuous change system,* an event causes a continuous change in the system over a period of time. These changes can be modeled by algebraic difference/differential equations and are coded in a subroutine which is integrated over time by the simulation package to obtain the system response.

In this paper we are concerned with the modeling of *discrete systems* and will illustrate the use of the *process* or *network* method and the *event* method, by modeling the assembly system outlined above.

IV. SIMAN and Cinema

Cinema is a graphical interface to the general-purpose simulation language SIMAN (SIMulation ANalysis). SIMAN and Cinema together provide an extremely flexible and graphically excellent modeling tool.

SIMAN is a FORTRAN-based simulation package which supports *network, discrete event,* and *continuous* simulation. It allows a programmer to fully describe a system using its command block and to set parameters for the system (Pegden, 1984). For the SIMAN model, the programmer creates two files, the system model file and the experimental frame file. These files are outlined below:

1. The system model file defines the *static* and *dynamic characteristics of the system*. For example, a part enters an input buffer for a workstation and waits there until the station becomes free. When it becomes available, the part seizes the station and waits there while it is being operated on. Once completed, it releases the station and enters the output queue. The part requests a transporter to bring it to the next station and waits in the output buffer until it arrives. This example could be represented by a model file using the block commands in Table 1.

As well as these commands, there is a selection of other commands which can be used to model the complex decision logic of manufacturing and assembly systems. In an extreme case, where SIMAN cannot model a given situation with the command blocks, a user-written FORTRAN subroutine can be called from within the program.

2. The experimental frame file defines the *experimental conditions* under which the model is run. For a given model file there can be many experimental

TABLE 1
The SIMAN Code Representing a Workstation

	A part:
QUEUE, 1;	Enters the input queue and waits
SEIZE: STATION_1;	Seizes the station when it becomes free
DELAY: 10;	Waits for ten seconds
RELEASE: STATION_1;	Releases the station
QUEUE, 2;	Enters the output buffer
REQUEST: TRANSPORTER	Requests a transporter to transfer the part to the next station

frames resulting in many sets of output data. The following is a list of possible types of information that might be found in an experimental frame:

- The number of stations and transporters
- The distances between points
- The statistics to collect
- The variables used in the model file
- The run length and number of runs for the simulation

The two files are linked together to produce one program. This program can be run on its own or in association with the Cinema graphics. At the end of a simulation run, the simulation results are sent to either a data file or the screen.

The basic construct used in creating animation within Cinema is the animation layout. A layout is defined as a combination of graphical objects which represent the system being modeled (Anonymous, 1984). Objects within a layout can be one of two types, static or dynamic. Cinema's animation can be described as the dynamic display of objects that change size, location, color, or shape on a static background.

The static objects form the layout background and represent the portion of the layout that does not change during animation. In a manufacturing system, this might correspond to the walls, aisles, or markings for transporter paths of the system being modeled. The static components of the layout are constructed using a set of elementary computer-aided drawing functions that allow the user to quickly add both graphics and text to the background.

The dynamic components incorporated into Cinema's layout are directly related to specific modeling elements of the SIMAN model. Each dynamic object added to the animation layout is associated with part of the system represented by the SIMAN model. Cinema can animate

- Entities/parts traveling through a system and the change in state as they are operated on. This is achieved by associating the individual entity icons with different attribute values associated with the part in the SIMAN model. During simulation of the SIMAN model, the attribute value of an entity may change. To reflect this change, the icon which represents the entity may also change.

• A queue which is positioned on the layout is associated with the corresponding queue used in the SIMAN model. Parts enter the queue and are displayed side by side until they are ready to leave.

• The changing states of a resource/station. The station is positioned on the layout and is associated with a specific resource in the model file. A station can have four different states: busy, idle, preempt, and inactive. For each of these states, the user draws a separate icon and as the station changes status in the SIMAN program, Cinema changes the icon shown on the screen. This can give the resource a sense of movement as well as highlighting its states.

• The transfer of parts through the system. There are three ways in which a part can be transferred: by transporter, conveyer, or direct route. In the SIMAN program and the Cinema layout, the transfer method is defined and the paths it travels described.

• Finally, Cinema can represent dynamic variables on the screen during simulation. It offers a selection of statistic variables used by SIMAN and allows the user to position the variable anywhere on the layout. This can be useful for monitoring resource utilizations, throughput, or any other statistic that may be required.

During the simulation run, Cinema offer special features to control and monitor the simulation. Some of these features are

• Control over the speed of the animation up to a maximum speed dictated by the hardware.

• A snapshot facility which will save the state of the simulation at the time the snapshot is taken. This saves graphics and system statistics.

• The option for a user to stop, pause, and/or reinitialize the simulation model at any stage.

We have used the IBM PC version of Cinema in our work.

A. THE PROCESS OF BUILDING THE MODEL USING SIMAN AND CINEMA

As Cinema is a graphical package which is interfaced with a SIMAN program, the first step in developing the simulation model is to create a SIMAN program. Using the details of that program, a graphical layout of the model can by created in Cinema.

SIMAN can model a system using any combination of the modeling techniques mentioned above, though in modeling the assembly lines for the kitchen appliance, the *process* method was sufficient. To describe how to use all of the various modeling possibilities in the language is outside the scope of this paper. We will limit our discussion to those features of the *process*-oriented simulation approach used in modeling the Bosch assembly line described earlier.

In SIMAN there are 10 main command blocks types. A block is a SIMAN command. The blocks, QUEUE, STATION, BRANCH, PICKQ, QPICK, SELECT, and MATCH perform a single function, whereas the blocks OPERATION, HOLD, and TRANSFER are block types and can perform many functions under different block names. To illustrate the capabilities of the SIMAN language, the functions of the block types used to model the kitchen appliance assembly line are now discussed in more detail:

- The OPERATION block is used to model a wide range of processes where an operation changes the status of the system or the value of a statistic.
- The HOLD block is used to model situations in which the movement of an entity or part is delayed based on the status of the system. When an appropriate operation occurs the part may continue.
- The TRANSFER block is used to model transfers between stations via material handling systems.
- The QUEUE block provides a waiting space for entities which are delayed by a HOLD or MATCH block.
- The STATION block defines the points between which parts are transferred by the TRANSFER block.
- The BRANCH block models stochastic and deterministic branching of entities to different parts of the model.
- The MATCH block delays entities in a set of preceding QUEUE blocks until entities with a similar characteristic, specified by its attribute value, reside in each queue.

These block commands are used together to develop a model of a system. As it is difficult to understand how these work in practice without seeing them in use, part of the file which represents the kitchen appliance assembly lines to be modeled is presented in Table 2. All these commands are part of the model file. To complement this file, there is an experimental frame file which declares the resources, transporters, and statistics to be collected.

The development of a graphical interface to the SIMAN program using Cinema is a relatively simple task. An icon representing each of the dynamic components to be animated is drawn. Cinema has a library facility which allows the user to create a library of icons which can be used with any model. The icons become part of the graphical layout when they are copied from a library and positioned on the layout. There are four icon libraries: resource, entity, transporter, and global symbols.

Using a well-defined procedure, a library for each icon type is created. The next task is to add the icons to the graphical layout. Before adding the icons,

TABLE 2
The SIMAN Code Representing Part of the Kitchen Appliance Assembly Line Model

	CREATE, 1: 15.0, 50;	OPERATION	Creates one entity every 15 seconds. There can be no more than 50 parts in the system at any one time
	ASSIGN: A(1) = 1;	OPERATION	Assign to the entity the value one on its first entity. Attribute one is used by Cinema to determine the entity icon
Queue_1	QUEUE, 1;	QUEUE	This queue holds unkitted parts until the kitter is available
	SEIZE: KITTER;	HOLD	This block holds entities in the previous queue until the kitter is available to be seized
	DELAY: 18.81;	OPERATION	Once the kitter has been seized the part is delayed for 18.81 s while it is being operated on
	RELEASE: KITTER;	OPERATION	The kitter is released and is now available to be reseized
	ASSIGN: M=1;	OPERATION	Assigning M = 1 defines the station at which the part is, before it is transported
	ASSIGN: A(1) = 2;	OPERATION	The entity icon used by Cinema is changed to the second icon
	QUEUE, 2;	QUEUE	The entity waits at queue 2 until it can access the conveyer
	ACCESS: BELT1, 10;	HOLD	The access block holds the part in the previous queue until it can access the conveyer BELT1
	CONVEY: BELT1, 2;	TRANSFER	The convey block transfers the part to station 2, using conveyer logic
	STATION, 2,	STATION	This block defines a place in the program which represents a point in the actual system
	BRANCH, 1: IF, NQ(3).LE.1, Q3: ELSE, NQ;	BRANCH	The branch block checks the size of the third queue. If there is a place in the queue the part is routed there. If not, the part is routed elsewhere

a background or static layout should be created. The background layout serves two purposes. First, it illustrates the connections between resource icons and the travel paths and, second, it enhances the effect of the icons and makes the layout more realistic. Figure 3 illustrates the layout used to model the kitchen appliance assembly lines. Once a model in SIMAN is developed and an associated graphical layout created, the animation is ready to be run.

B. VALIDATION OF THE SIMAN MODEL

The next stage in the simulation process is the validation of the developed model. Validation is a two-stage process—verification and validation proper. The difference between validation and verification is brought out by Schlesinger (1974) "*Verification* refers to the process of confirming that the conceptual model has been correctly translated into an operational program, and the calculations made by this program utilize the correct input data. *Validation* refers to the process of confirming that the conceptual model is applicable or useful by demonstrating an acceptable correspondence between the computation results of the model and actual data." Figure 4 (Browne and Davies, 1984) illustrates the distinction. The kitchen appliance assembly line model was built using many different simulation packages. Each model was verified individually. In verifying

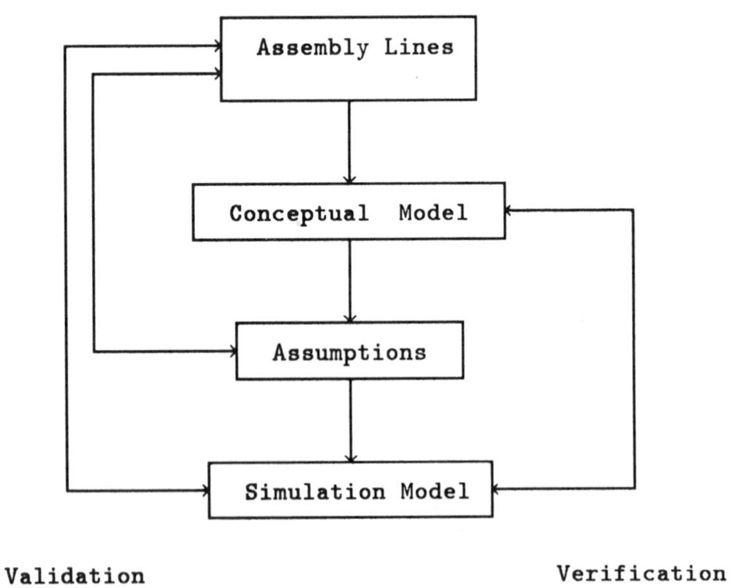

Figure 4. The validation and verification of a simulation model.

the model, the internal logic of the model was checked. In all packages, the final model performed as it was expected to, taking into account its limitations.

The actual validation of the models was achieved by comparing the SIMAN/Cinema model with the real system and in turn comparing the other models with it. To validate the SIMAN model, three engineers who were working on the kitchen appliance assembly line examined the simulation model. First, the assumptions on which the model was based were reviewed and accepted by the engineers. The assumptions had been presented to the engineers prior to development of the model.

The next stage was to allow them to monitor the animation and compare the simulation results with the real system. The output from the simulation model was similar to the actual system, specifically

• Workstation utilizations were comparable. On the feeder line the operator at the fifth assembly station was starved of work. In the real system they had to remove this operator for part of each day. Also, an active bottleneck on the lower line was highlighted by the simulation model.

• The throughput of the model corresponded favorably to the throughput of the real system.

• The flow of parts displayed by animation in Cinema highlighted the problem of clustering, where groups of parts cluster together and travel around the assembly line. This problem was also found in the real system.

As the initial SIMAN model favorably represented the real system, it could be used to validate the other models. Having gained confidence in the model, the problems highlighted by the simulation and which had actually arisen in the real system can then be addressed by experimenting with the model. It was at this stage that the model was most useful. Before looking at the experiments carried out on the model, the type of statistical results output by SIMAN are outlined.

C. SIMULATION RESULTS FROM THE SIMAN MODEL

In the SIMAN experimental file, all statistics that are to be recorded are specified and in some cases linked to the model file. In the model of the kitchen appliance assembly line there were three types of statistics recorded:

• Counters record the number of entities/parts that pass different points in the system. By comparing a counter at the entry point to the system and the exit point, a work-in-progress level can be calculated.

• Tallies record statistics such as *time in system,* and *interarrival* times.

TABLE 3
Resource Utilization Levels

		Discrete change variables				
Number	Identifier	Average	Standard deviation	Minimum value	Maximum value	Time period
1	MOTKIT UTIL.	0.87	0.33	0.00	1.00	28800.00
2	ASSEMB. 1 UTIL.	1.00	0.03	0.00	1.00	28800.00
3	ASSEMB. 2 UTIL.	1.00	0.06	0.00	1.00	28800.00
4	ASSEMB. 3 UTIL.	1.00	0.08	0.00	1.00	28800.00
5	ASSEMB. 4 UTIL.	0.99	0.09	0.00	1.00	28800.00
6	ASSEMB. 5 UTIL.	0.62	0.49	0.00	1.00	28800.00
7	RUN IN UTIL.	0.92	0.27	0.00	1.00	28800.00
8	TEST UTIL.	0.92	0.27	0.00	1.00	28800.00

• Discrete statistics are recorded on any of the system components, such as resources, queues, transporters, and/or conveyors. The user specifies the components on which statistics should be collected.

Discrete statistics and tallies are presented in tabular form indicating the average value, the standard deviation, and the maximum and minimum values. Using these statistics any information concerning the system can be obtained. Table 3 lists some of the discrete statistics computed during simulation of the assembly line model. Any of the statistics collected by SIMAN can also used by Cinema. Cinema can display the statistics in both numerical or graphical form.

D. EXPERIMENTATION WITH THE KITCHEN APPLIANCE ASSEMBLY LINE MODEL

Before experimenting with the model, it was important to understand the problems of the actual line. The simulation results highlighted two main problem areas on the feeder line. It was with these in mind that the experiments were planned. The problem areas were as follows:

• As all of the motor assembly operators were being paid through a bonus plan, the operators expect that they would be constantly supplied with work. The simulation highlighted the fact that the last assembly station was being starved of kitted pallets. In fact, there are no pallets available 38% of the time.

• Another problem area was the connection between the feeder line and the main assembly line. The finished motors were fed in a JIT manner to the lower line. Simulation of the assembly lines highlighted the fact that a large number of parts queue up at the end of the feeder line waiting to leave. The queue of parts sometimes stretches back as far as the TEST station, causing blockages. The queue of parts built up because the main assembly line operated at at different rate to the feeder line.

The line engineers on the kitchen appliance assembly line also had a problem. They were unsure as to the number of pallets that would be required on each line. In an attempt to solve these problems, some experiments were carried out on the model. Four of these experiments are briefly outlined below.

• It can be seen in the simulation results that one of the motor assembly operators was starved of work. It was decided to try a *different dispatching rule which would not discriminate between these stations*. The new dispatching rule ensured an even supply of pallets to each assembly station by transferring pallets to the stations in a cyclic order. The changes were implemented and the simulation model was rerun. The utilization of the last motor assembly station improved dramatically to 92% but this was at the expense of all the other stations whose utilizations were reduced to a similar figure. Other dispatching rules were tested, but they simply reallocated the work between the operators. It was also noted that the queue at the end of the feeder line remained unchanged.

• After trying different scheduling techniques and not improving the system's performance, the focus of attention turned to the connection between the two assembly lines. It was decided to create a buffer stock of motors at the end of the line and not to feed the lower line in a JIT manner. All empty pallets would be recirculated back to the kitter. When this change was made to the model and simulated, it was found that all the assembly operators were kept supplied with kitted pallets except for the last operator who was idle for 5% of the time. The dispatching rule used above was again tried, and this time all the operators had sufficient work, but now the line was overproducing. That is, at the end of each shift there were 240 unused motors.

• The next experiment had to take into account some external information. There is a second shift which produces the same appliance but because of supervision problems it does not use the Bosch assembly lines. This shift produces between 200 and 300 appliances, depending on demand. It was suggested that the day shift could supply the motors for the evening shift quota. The problem was then to see how much overtime would be required. The model was again changed, this time allowing the simulation run to continue until the evening shift's quota was reached as well as the normal day requirements. It was found that, in general, overtime was unnecessary but at high demand (300 appliances), overtime of 1.66 h per person per week was necessary.

• To answer the question about the number of pallets required for each line, the number of pallets was varied for both lines. Originally the engineers expected to have 50 pallets on the feeder line and 40 on the main assembly line. The problem was to ensure that there were always enough pallets available for kitting. On the other hand, there could not be too many pallets on the line, as blocking would occur. For each simulation run, the queues before bottlenecks in the system and the kitting stations were examined. It was found that there was no need for 50 pallets on the feeder line, as many pallets spent considerable time in the carousel. The simulation results suggested that the "optimum" number of pallets was 39. It was also noted that the number of pallets needed was directly related to the speed of the conveyor system.

On the main assembly line, it appeared that the number of pallets was too small. Even though queues did build up, some stations were starved of work because there were no pallets available. The "optimum" number suggested for the main line by the simulation model was 46.

The preceding experiments are just a sample of the those carried out and demonstrate how simulation can be used *to support the detailed analysis and design of flexible assembly systems.*

E. FEATURES OF THE SIMAN SOFTWARE

SIMAN is extremely flexible and can model a system using any combination of the simulation modeling techniques identified earlier. Its strength lies in its supporting software, in particular

- BLOCKS, a graphical tool which can be used to create a model file by selecting the appropriate command block from a pull-down menu and placing it in the program. The user is prompted for all relevant parameters of the block.
- ELEMENTS, which operates in a similar manner to BLOCKS to create the experimental file.
- Cinema, which was discussed above, is a graphical front end to SIMAN. Cinema animates the simulation model using a graphical layout specified by the user.
- OUTPT, which takes specified output from SIMAN and presents the results in histogram form.

SIMAN tends to offer a greater selection of *network* command blocks than other simulation languages. Therefore, it can model more complex systems using its *network/process* modeling.

V. SLAM II

SLAM II, the *Simulation Language for Alternative Modeling,* is a flexible FORTRAN-based simulation language. It has the power to model a system using any of the three simulation modeling approaches, *process/network, discrete event,* and *continuous* separately, or it can use any combination of the approaches in one model (Pritsker, 1984). SLAM II is particularly strong in the area of event-oriented modeling.

A simulation model is developed using SLAM II special modeling symbols. These symbols are entered into a data file and are called from a FORTRAN program which includes the SLAM II routines and the user's own subprograms. As a model becomes more detailed and complex, *discrete event* and *continuous* modeling features can be added. The format of a SLAM II model allows any SLAM II symbol or any predefined user-written subprogram to be added to the program without recompiling the initial program. As a result, a model may be modified and resimulated without great effort.

SLAM II automatically collects standard statistical information after each simulation run and will record any other type of statistics that the programmer specifies. It is only after analyzing these results that the model can be verified.

A. THE PROCESS OF BUILDING THE MODEL USING SLAM II

A simulation model normally begins with a *flow diagram* or *network file which portrays the flow of entities (people, parts, or information) through a system*. A SLAM II network model is made up of nodes at which operations are performed. In the program, these nodes are represented by the SLAM II symbols. There are 20 node types with functions such as entering or exiting the system, seizing or freeing a resource/workstation, changing variable values, collecting statistics, and stopping or starting the entity flow based on system conditions. Nodes can be connected directly or by branches which define the routings of entities based on certain conditions. Branches may be deterministic, stochastic, or based on system variables. Time delays may represent processing times, travel times, and/or waiting times.

As a model grows in detail, the network functions may not be able to fully model the system. For this reason, SLAM II provides easy access to user-written FORTRAN subroutines for more complex *discrete event* modeling. In the SLAM II model of the kitchen appliance assembly lines, the transfer of parts by conveyor was modeled by a FORTRAN subroutine. To illustrate the type of network commands used, a listing of the symbols and a brief definition of each is given below.

- An ACTIVITY statement specifies a time delay and/or entity routing.
- Variables and entity attributes are ASSIGNed values, in order to control the flow of the entities/parts through the system.
- The COLCT node defines special statistics that should be recorded during the simulation run.

TABLE 4
SLAM II Code Representing the Connection between the Two Assembly Lines

	AWAIT(14),TEST/1,1;	A part arrives from the previous station and waits for the test station to become free.
	ACT/8,20.0;	When the station is free the part is held for 20.0.
	FREE,TEST/1,1;	After waiting, the station is freed and the part continues.
E15	ASSIGN,ATRIB(4)=14.0, ATRIB(5)=15.0,1;	ATRIB(4) is assigned the number of the present position, and ATRIB(5) is assigned the number of the next station.
	EVENT,1,1;	The event node calls a user-written subprogram which checks the size of the next queue. If there is room, ATRIB(6) is assigned the number 1.0.
	ACT,0.5,ATRIB(6).EQ.0.0,E15;	If there is no room in queue 15, the entity tries again.
	ACT,USERF(1),ATRIB(6).EQ.1.0;	If there is room, the part is transferred to the next station. The transferring logic is coded in subroutine USERF(1).
	ACT,,,Q1;	Parts are transferred to the queue labeled Q1.
Q1	QUEUE(1),,,,MATC	In queue 1 the parts wait until they can be matched with another part in queue 2.
	CREATE,15.0,,,50,1;	One new pallet for the main assembly line is created every 15 s up to a maximum of 50 pallets.
	ACT,,,Q2	Empty pallets are transferred to the queue labeled Q2.
Q2	QUEUE(2),,,,MATC	The empty pallets wait in queue 2 until it can be matched with a part in queue 1.
MATC	MATCH,,Q1/E1,Q2/P16	The match node holds the parts in the preceding queues until a pair can be matched. When they are matched, the parts in queue 1 are sent to a node labeled E1, and similarly the pallets in queue 2 are sent to P16.

- The CREATE node creates entities based on a specific arrival pattern.
- The GOON node does not carry out a specific operation, but acts as a step between two other nodes.
- The MATCH statement holds entities in preceding QUEUE nodes, until entities in a number of queues share the same attribute value and therefore can be matched.
- The QUEUE node holds entities until a server becomes available.
- The RESOURCE node defines a resource (workstation) and its capacity.
- The EVENT node transfers control to a user-written subprogram.
- The AWAIT node holds entities until a resource is available or a gate is opened.
- The FREE node frees a resource and makes it available for reallocation.

A network model of a system can be built and user-written routines called using the preceding command nodes. To show how a SLAM II program is structured, part of the SLAM II program which models the kitchen appliance assembly lines is presented in Table 4. There is an explanation of each command on the right side of the program. This extract from the SLAM II file gives a quick insight into the workings of SLAM II.

B. VERIFICATION OF THE SLAM II MODEL

The SLAM II model was both verified and validated by comparing it with the SIMAN model. There was no practical difference in the simulation results, even though the SLAM II model used *event*-based subroutines to model the conveyor system and the SIMAN model used a set of its own block commands to model the conveyors.

C. SIMULATION RESULTS FROM THE SLAM II MODEL

SLAM II automatically records and outputs statistics for each queue, resource, and activity. Additional statistics may be specified, such as interarrival times at certain points and the time it takes a part to travel between points. All statistics are presented in a tabular form and for each type of statistic there is a separate table. For example, there is a stable for queues and a table for resources. Table 5 displays a selection of resource statistics taken from the simulation results for the kitchen appliance assembly line model. There are similar tables of statistics for files/queues, activities, and any other specified entities. SLAM II outputs two types of simulation reports: intermediate reports which are recorded and stored at different stages during the simulation, and the final summary report.

TABLE 5
Resource Statistics Outputed by SLAM II

Resource statistics						
Resource number	Resource label	Current capacity	Average utilization	Standard deviation	Maximum utilization	Current utilization
1	KITTER	1	0.87	0.332	1	1
2	MOT_ASS_1	1	1.00	0.033	1	1
3	MOT_ASS_2	1	1.00	0.058	1	1
4	MOT_ASS_3	1	0.99	0.079	1	1
5	MOT_ASS_4	1	0.99	0.098	1	1
6	MOT_ASS_5	1	0.63	0.483	1	1
7	RUN-IN	1	0.92	0.274	1	1
8	AUTO-TEST	1	0.92	0.276	1	1

D. FEATURES OF THE SLAM II SOFTWARE

SLAM II has many features and is extremely flexible. It can model a system using any combination of modeling techniques. The kitchen appliance assembly line model highlights two features of SLAM II:

- SLAM II has a very good blocking feature, where an entity/part is not allowed to proceed to the next station or queue unless it can be accommodated. This is true at all stages of the model and simplifies the process of modeling blockages, which are common in flexible assembly systems.
- There is no necessity to recompile a SLAM II code when changes are made to the network model, unless some changes are made within a FORTRAN subroutine.

Finally, SLAM II also has supporting animation software. TESS, the extended simulation system, is a software tool which supports the building of graphical models using the SLAM II language and includes both an animation and a graphical report facility (Standridge, 1984).

VI. MAST

MAST is a data-driven general simulator written in the programming language FORTRAN (Anonymous, 1986b). MAST has been designed to simulate flexible manufacturing systems and therefore is limited in the range of systems it can model. MAST is a collective name for three interconnected packages: MAST—manufacturing system design tool, SPAR—system planning of aggregate requirements, and BEAM—background and enhanced animation for MAST.

SPAR allows the user to input the production requirements and the process plans for all the parts to be processed by the system. Using static capacity analysis of the system described, the system's resource requirements can be determined in terms of process stations, transportation, and pallets. If the model fails to meet the resource requirements, SPAR identifies where extra resources may be necessary. After the capacity analysis, SPAR generates data input suitable for MAST (Anonymous, 1986c). SPAR therefore provides a first rough-cut capacity check of the system under study.

MAST is a general simulation model oriented toward the modeling of FMS. A particular FMS is described by the data entered in SPAR. MAST offers a wide range of control strategies and therefore a realistic representation of system control can be modeled (Anonymous, 1986b). If there is no suitable control strategy available, the user may write a FORTRAN subroutine which represents the required control strategy and link this subroutine with the MAST model. MAST creates three types of output:

- Statistical information concerning the parts, stations, transporters, and pallets.
- A basic analysis of the statistical output using a simple expert system, which suggests possible strategies for the redesign of the system, where appropriate.
- An event file, recording part and transporter movements, as well as station and pallet statistics. This file is read by the BEAM software which reproduces graphically the events that occurred during simulation.

BEAM provides the capability to generate a graphical representation of the model using both the data supplied by SPAR, and its own user interface (Anonymous, 1986a). The data BEAM requires include the type of icon which will represent a station and the position of that station. There is a limited number of icons available and the input and output buffers are placed in a standard position on each side of the appropriate icon.

After the system has been described and the simulation model is run, BEAM animates the events that occurred during the simulation and provides graphical displays of the performance statistics.

A. THE PROCESS OF BUILDING THE MODEL USING MAST, SPAR, AND BEAM

Before SPAR can perform capacity analysis, an outline of the manufacturing system must be entered. There are four types of information required to describe the system: the planning horizon, the parts' description, the stations' description, and the transporters' description. All data are entered through SPAR's interface, which prompts the user for the information. SPAR's data requirements can be summarized as follows:

• The planning horizon is the length of time over which the capacity of the system is checked.

• The part description menu has a field for each data entry. The part description includes, for each part type, the part identification, the production requirements, the number of parts that can fit on a pallet, a list of operations that have to be carried out on the part, the time (per pallet) to carry out that operation, and a storage factor. The storage factor is used by SPAR when it carries out the capacity analysis. It can be defined as the time spent by a part in storage relative to the time spent being processed. Figure 5 illustrates the screen that describes the motor assembly and the operations it undergoes.

• SPAR takes the full list of operations entered for all parts and assumes that for every operation type there is one station type. In the station description menu, the user specifies the number of stations that will carry out each operation and the time to exchange pallets at those stations. Figure 6 illustrates a screen listing the operations and stations for the feeder line.

• The transporter description menu prompts the user for the number of transporters and the expected cycle time for each transporter. Figure 7 illustrates the screen that defines the conveyors. By specifying zero transporters, MAST assumes that a conveyor system is being used. The average assignment time is equivalent to the average expected time for a part to travel between stations.

• Most of the information necessary to describe the system has, at this stage, been entered. All that remains is to enter the number of pallets in the

```
System Planning for Aggregate Requirements

PART ID(10 CHAR)         MOTOR
PRODUCTION REQUIRMENT    1250
PARTS PER PALLET         1

ENTER OPERATION DESCRIPTION(10 CHAR) AND OPERATION TIME
Type DELETE To Erase Operation From List

 1        OPERATION ID: KITMOT          TIME:    0.31
 2        OPERATION ID: ASSEMBLY        TIME:    1.67
 3        OPERATION ID: RUN-IN          TIME:    0.33
 4        OPERATION ID: TEST            TIME:    0.33
 5        OPERATION ID: KITCAB          TIME:    0.38

OPERATION NUMBER TO CORRECT :?
```

Figure 5. SPAR's part description.

```
System Planning for Aggregate Requirements

ENTER NUMBER OF STATIONS WHICH ARE AVAILABLE FOR EACH TYPE OF OPERATION

OPERATION ID: KITMOT      NUMBER OF STATIONS: 1     PALLET EXCHANGE TIME: 0
OPERATION ID: ASSEMBLY    NUMBER OF STATIONS: 5     PALLET EXCHANGE TIME: 0
OPERATION ID: RUN-IN      NUMBER OF STATIONS: 1     PALLET EXCHANGE TIME: 0
OPERATION ID: TEST        NUMBER OF STATIONS: 1     PALLET EXCHANGE TIME: 0
OPERATION ID: KITCAB      NUMBER OF STATIONS: 1     PALLET EXCHANGE TIME: 0
```

Figure 6. SPAR's station description.

system, to describe the track layout, to enter station buffer sizes, and to set the transporter speed.

The layout is based on the concept of decision points. A decision point is a point between which parts can travel and at which a decision can be made. The points are placed strategically around the model and there is a decision point at every input buffer and output to a station. Also there is a decision point at the point where two or more tracks join or a track divides. Therefore, to describe a track layout it is necessary to supply SPAR with details of which points are directly connected, and the distances between them.

Finally, all stations are assigned a decision point at their input and output buffers. In the same menu, the buffer sizes are entered for each station. Figure 8 illustrates the data entered for the KIT-Motor station.

SPAR's capacity analysis is simply the aggregation of operation lines and part quantities to determine if, over the planning horizon, the production requirement can be reached. It is an effective rough-cut check to determine if the system is capable of achieving its projected output. It produces four reports—a production requirements summary, a station utilization report, a transporter utilization report, and a statement of pallet requirements. Figure 9 shows the station utilization report for the feeder line. The analysis of the motor assembly stations proceeds as follows: There are five assembly stations which have to process

```
System Planning for Aggregate Requirements

TRANSPORTER DESCRIPTION

NUMBER OF TRANSPORTERS  : 0

AVERAGE ASSIGNMENT TIME : 0.06
```

Figure 7. SPAR's transporter description.

```
System Planning for Aggregate Requirements

KITMOT STATION

STATION NUMBER    1

1. Number of on-shuttle queue positions         :? 2
2. Number of off-shuttle queue positions        :? 1
3. Decision Point where parts are delivered     :? 1
4. Decision Point where parts are picked up     :? 2
```

Figure 8. SPAR's description of the association between queues, decision points, and a workstation.

1250 parts between them, in 480 min. Therefore the stations are available for 480 × 5 = 2400 min. They are required for 1250 × operation time (1.66) min = 2082.5 min. From these times, the station utilizations and slack can be calculated. Similar calculations are carried out for transporter and pallet requirements. If the model does not meet the requirements, SPAR suggests possible changes to the system design.

The data required by BEAM to describe a graphical layout are the position of each station on the graphical layout, the size of each queue for graphical purposes, and the identification of the decision points through which the conveyor/transporter will travel. Figure 10 illustrates the layout of the kitchen appliance represented by BEAM.

Finally, before running the simulation, MAST allows the user to choose a set of control strategies to manage the flow of parts through the system. The strategies are used to define the answers to certain control questions, including

- When should a part be introduced into the system?
- For each station, which parts should be taken from the input buffer when there is more than one part available?
- After an operation is finished, which operation should be carried out next?

STATION CAPACITY ANALYSIS (SPAR)

STATION TYPE	NUMBER AVAILABLE	TOTAL TIME AVAILABLE	TOTAL TIME REQUIRED	STATION UTILIZATION	SLACK PER STATION
KITMOT	1	480.0	391.9	81.6	88.0
ASSEMBLY	5	2400.0	2082.5	86.8	63.0
RUN-IN	1	480.0	416.6	86.8	63.0
TEST	1	480.0	416.6	86.8	63.0
KITCAB	1	480.0	471.3	98.2	8.0

Figure 9. SPAR's capacity analysis of the workstations.

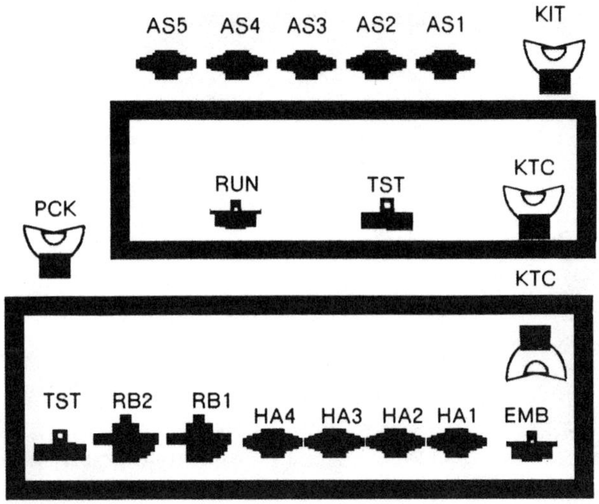

Figure 10. BEAM's representation of the kitchen appliance assembly lines.

- Knowing the next operation, to which station should the part be sent?
- Which transporter should transport the part to the next station?
- How should transporter movement be controlled?

For each of these questions, there are typically five possible answers. If none of the answers is appropriate, it is possible to write a FORTRAN subroutine which models the required control strategy and which can then be to linked to the MAST model.

B. VERIFICATION OF THE MAST MODEL

Perhaps the most important feature of MAST is that it is designed as a tool to facilitate the design of flexible manufacturing systems, in particular flexible machining systems. As such it was not really appropriate to the kitchen appliance assembly line. However, it is clear that in the context of machining systems, MAST is a very powerful and easy to use simulation package.

C. SIMULATION RESULTS FROM THE MAST MODEL

MAST produces two types of simulation statistics, a graphical representation of the different elements of the simulation results and a listing of the results in tabular form. The graphical results can be viewed during the animation and are updated as the animation progresses. Within the animation environment

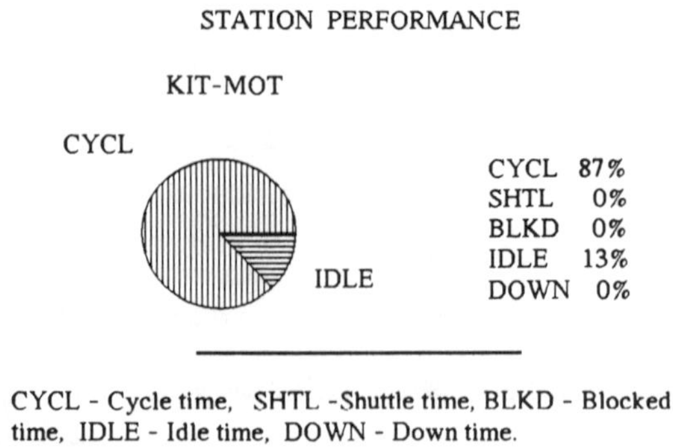

Figure 11. Graphical output from MAST—station performance.

the user has access to bar charts, pie charts, and trend graphs which represent the part performance, the production performance, the station performance, and the transporter performance. Figure 11 illustrates a pie chart which represents the performance of one of the motor assembly stations on the feeder line.

The second type of simulation results appear in tabular form. MAST produces two sets of statistical results, intermediate results which are collected at user-specified time intervals and a summary report which is recorded at the end of each simulation run. As in most simulation systems, detailed statistics are collected on each component of the model. MAST produces reports on

- Production levels and time-in-system statistics
- Part performance statistics, indicating where a part spends its time

TABLE 6
Station Performance Statistics for the Feeder Line

			Station performance summary							
Station number	Total cycle time	(%)	On/off shuttle time	(%)	Part blocked on work table time	(%)	Idle time	(%)	Down time	(%)
1	419.0	(87.3)	0.0	(0.0)	0.0	(0.0)	61.0	(12.7)	0.0	(0.0)
2	463.2	(96.5)	0.0	(0.0)	0.0	(0.0)	16.8	(3.5)	0.0	(0.0)
3	461.2	(96.2)	0.0	(0.0)	0.0	(0.0)	18.2	(3.8)	0.0	(0.0)
4	432.5	(90.1)	0.0	(0.0)	0.0	(0.0)	47.5	(9.9)	0.0	(0.0)
5	427.7	(89.1)	0.0	(0.0)	0.0	(0.0)	52.3	(10.9)	0.0	(0.0)
6	427.2	(89.1)	0.0	(0.0)	0.0	(0.0)	52.3	(10.9)	0.0	(0.0)
7	442.6	(92.7)	0.0	(0.0)	0.0	(0.0)	35.0	(7.3)	0.0	(0.0)
8	442.6	(92.7)	0.0	(0.0)	16.3	(3.3)	19.2	(4.0)	0.0	(0.0)
9	480.0	(100.0)	0.0	(0.0)	0.0	(0.0)	0.0	(0.0)	0.0	(0.0)

• Pallet information, indicating the number of pallets used and the maximum number needed
• Station performance statistics, indicating the station's busy time, loading time, blockage time, idle time, and downtime
• Queue performance statistics, which show the average and maximum size of the queues
• Cart or transporter performance statistics, which show similar statistics to those produced for the stations

Table 6 illustrates the station performance statistics for the first nine stations on the feeder line.

D. FEATURES OF THE MAST PACKAGE

MAST is unique among the simulation packages which are described in this paper from two points of view:

• The MAST package includes SPAR, which facilitates capacity analysis of the proposed FMS prior to detailed simulation. This is a good feature since it allows the user to determine if the proposed solution is reasonably acceptable prior to the detailed simulation work which MAST itself supports.
• MAST includes an "expert system" which analyzes the results of a simulation run and, where appropriate, suggests to the user modifications to the FMS design.

MAST offers a very comprehensive error-checking facility in terms of the syntax and completeness of the data input to it. Further, the animation facility within MAST is achieved by retaining a trace file of all the events which took place during a simulation run. This allows the animation to be rerun without rerunning the simulation model.

VII. SIMFACTORY

SIMFACTORY is a data-driven simulation model of a generic factory. It allows the user to describe the important components and constraints of the factory, using both data input and graphical tools. Using this information, SIMFACTORY constructs a model of the factory using the simulation language SIMSCRIPT II.5 (Anonymous, 1987b). The simulation can be run with animation of varying degrees (e.g., the animation of part movement may be turned on or off). Also the animation may be customized to individual users needs. No programming expertise is required to create a SIMFACTORY model. Everything is accomplished by entering tables of data which describe the manufacturing system to be simulated. There are four main programs that make up the SIMFACTORY package:

- The SIMFACTORY model
- The user interface
- The layout editor
- The icon editor

The SIMFACTORY model is a general-purpose manufacturing/assembly simulation model. It is based on the premise that all manufacturing and assembly systems contain the same basic "building blocks": stations, queues, transporters, resources, maintenance, and so on. A model of one factory is distinguished from another by the ordering and content of the building blocks and the parameters used, for example, speed, capacity, and so on. Using the system description provided by the user interface and the layout editor, SIMFACTORY generates a model and simulates the system. It also provides a large range of reporting facilities which the user selects. The user interface is a system of interlocking menus used to describe new models, edit existing ones, run models, and examine the simulation output. The data entry is well organized and easily modified. The layout editor is a graphical tool which enables the user to create and edit the layout of a factory model. The layout is both a physical and logical representation of the actual factory layout. Machines, queues, and transporters are placed in the correct physical locations. Also, connections between the components are described graphically. The icon editor is a graphical editor which allows the user to create and edit the graphic icons which represent the components of his model. An icon can represent stations, queues, transporters, parts, and background shapes.

A. THE PROCESS OF BUILDING THE MODEL USING SIMFACTORY

In SIMFACTORY, the information required to describe a factory model is called a dataset. There is a specific menu which describes each subsystem of the dataset and in the case of the factory layout there is also a graphical tool. A complete dataset contains the following subsystems:

- A factory layout
- An options specification, which defines the run length and report features
- The set of all stations referenced by the layout
- The set of all queues referenced by the layout
- The set of all transporters referenced by the layout
- The set of process plans
- The set of all interruptions referenced by stations and transporters

To add or edit any part of the dataset, the user selects the appropriate item from the main menu and SIMFACTORY presents the appropriate environment.

1. Describing a Station Using SIMFACTORY

SIMFACTORY prompts the user to describe a set of stations for each model. It defines a station as a place at which parts are processed. A station can perform multiple operations and may be subject to interruptions. The information required to describe the kitting station on the feeder line (see Figure 12) is as follows:

• The *station name* which will be used in the layout. The kitting station is called KIT-Motor.

• The *icon* which will represent the station in the layout. This station uses an icon which is known as Packer.

• The *type of station*. There are four types of stations offered by SIMFACTORY. The kitting station is a "normal" station.

• The most important piece of information needed to describe the station, is a *list of the operations* carried out at the station. The kitting station performs one operation, i.e., it places the motor components on a pallet. This operation is known as KIT-Mot.

• The *efficiency factor* of the station. As two stations carrying out the same operations may operate at different rates, an efficiency factor is associated with each operation for each station.

```
SIMLAB——Assembly Line————————————————————————————————————————
status: Editing Station KIT-MOTOR
                         S T A T I O N   E D I T O R

 Interruptions (F5)

 Icon:  Packer          Id: 1           Type: normal        Capacity: 1

         Operations:  Name      Efficiency          Setup Time
        ┌──────────────────────────────────────────────────────────┐
        │             KIT-MOT   1.0                 .0000          │
        │                                                          │
        │                                                          │
        └──────────────────────────────────────────────────────────┘

 Sequencing Rules:     Station Inventory                 Downstream Domain
                    first by: FIFO                    first by: Fixed Order
                    then by:  FIFO                    then by:  (unspecified)

   SIMFACTORY               r1.53
```

Figure 12. A station description within SIMFACTORY.

• The *setup time* between operations. If a station carries out more than one operation there may be a need to change the setup of the station. Also, when a part first arrives at a station it may need to be set up.

• Operations performed at a station may require a *resource for setup or processing*. For example, in the case where an operator is required to change the settings at the station, he is considered to be a resource. A resource can be used to model operators, tools, fixtures, containers, or pallets.

• A station may experience two types of *interruptions,* random breakdowns and/or preventative maintenance.

• For each station, there are two *sequencing rules*. The first rule defines the order in which parts are processed. There are four possibilities: first in first out (FIFO), last in first out (LIFO), earliest due date (EDD), and latest due date (LDD).

When parts leave a station, there may be a number of downstream stations and/or queues to which it may travel. There are two rules governing the selection of a downstream station/queue. A part can either check each station/queue in a *fixed order* for an available place, or select the station/queue which has been available for the longest time, namely *first available*. In the case of fixed order, the order in which downstream places are checked depends on the order in which they are located.

```
SIMLAB----Assembly Line------------------------------------------------
status: Editing: queue KIT-QUEUE
                              Q U E U E   E D I T O R

 Icon:  input            Id:           Content: part           Capacity: 20

 Raw Materials:  | MOTOR-COM                                              |

        -sui-------------------------------------------------------------
        |
        |                R A W   M A T E R I A L   E D I T O R
        | Lead Time                        C(.0000)
        | Mean Time Between Arrival        C(18.0)
        | Mean Arrival Quantity            C(1.0)
        | Due Time                         C(.0000)
        |
        -----------------------------------------------------------------

 Sequencing Rules:     Station Inventory                 Downstream Domain
                    first by: FIFO                   first by: First Available
                    then by:  FIFO                   then by:  First Available


   SIMFACTORY              r1.53
```

Figure 13. A queue description within SIMFACTORY.

2. Describing a Queue Using SIMFACTORY

SIMFACTORY defines a queue as an area used to temporarily store parts or resources. A queue is described by its content type, capacity, and sequencing rules. To fully describe a queue (see Figure 13) the following fields must be filled:

• The *icon field*. Each queue can be represented on the layout by any of the defined queue icons.

• The *content type* of a queue indicates whether a queue holds parts or resources. A part is an item which is processed at stations, whereas resources are used to help set up a station or process a part. Parts and resources enter the system via queues. Parts enter the system as raw material according to an arrival pattern, and are stamped with the time they are due to be completed.

• The *capacity* of the queue defines the maximum number of parts which can reside in a queue at any one time.

• Finally, the *sequencing rules* operate in the same manner as those used by a station.

3. Describing a Transporter Using SIMFACTORY

SIMFACTORY defines a transporter as a method for moving parts through a factory. Transporters can be used to carry parts between pick-up and drop-off locations. To fully describe a transporter (see Figure 14) the following data must be entered:

• The *icon field:* As with stations and queues, a transporter can be represented on the layout by any of the defined transporter icons.

• The *transporter type:* There are two type of transporter, batch or conveyor. A batch transporter is an individual transporter which travels between the pick-up and drop-off points. A conveyor is a continuously moving unit which also moves parts from point to point but does not physically change location itself.

```
SIMLAB——Assembly Line————————————————————————————————
status: Editing: Transporter Converor
                        Q U E U E   E D I T O R

 Icon:  dummy           Id:             Speed: .20

 Type:   conveyor       Transfer:   conditional

 Conveyor Type:   accumulating              Part Capacity:  200


   SIMFACTORY                r1.53
```

Figure 14. A transporter description within SIMFACTORY.

• The *transfer mode:* A transporter can operate in two modes. It can either transfer parts to its next location without checking to see if there is sufficient room, that is, on unconditional transfer; or it can transfer a part only if there is an available place, conditional transfer. Unconditional transfers may cause blockages at the entry points to stations.

• The *transporter speed* defines the speed at which a transporter travels. The travel distances are defined in the layout.

• Finally, in the case of a conveyor its *capacity* is defined, that is, the maximum number of parts that can be fitted on the conveyor at any one time.

4. Describing a Process Plan Using SIMFACTORY

A process plan includes part routings and processing times. It consists of a set of input parts, a sequence of operations, and a set of output parts. Using the concept of multiple inputs and outputs, SIMFACTORY can model all the manufacturing operation types defined by Suranjan and Nof (see Suranjan *et al.*, 1985). To fully describe a process plan (see Figure 15), the following information is required:

SIMLAB——Assembly Line——
status: Editing: MOTOR-ASS

P R O C E S S   P L A N   E D I T O R

Inputs	Icon Mane	Quantity
MOTOR-COMP	white part	1

Operations	Processing Time Distribution
KIT-MOT	C(18.81)
ASSEMBLY	C(100.0)
RUN-IN	C(20.0)
TEST	C(20.0)
KIT-CAB	C(19.16)

Outputs	Quantity	Probability
CAB-COMP	1	1.0

SIMFACTORY r1.53

Figure 15. A process plan description within SIMFACTORY.

• The *inputs* to a process may be either raw material or work in progress. For each input the name, quantity, and correct icon must be specified.

• The *list of operations* which convert inputs onto outputs. As a station can perform multiple operations, SIMFACTORY can model the flexible routing of parts associated with flexible assembly system by checking the list of operations and selecting the next station. Also, there is a time distribution associated with each operation and where appropriate a random number stream.

• The *outputs* from a process plan represent value added to work in progress, final parts, or by-products. These outputs can act as inputs to process plan.

5. Creating a Graphical Layout Using SIMFACTORY

The individual elements of a factory from a simulation viewpoint are comprised of stations, queues, and transporters. Each element is defined outside the layout environment. These elements are then combined in the factory layout to produce a working model of a complete factory. The factory layout serves three purposes:

1. It selects from the station, queue, and transporter libraries those individual elements which will be included in the factory layout.
2. It defines the physical arrangement of the selected elements.
3. It defines how those elements are connected.

The first and second functions are relatively simple. The third function is probably the most important aspect of the factory layout since work flow through a factory is determined in large part by the way in which elements are connected. A connection in SIMFACTORY defines how parts move through the factory. In order to describe how SIMFACTORY models physical connections, it is necessary to explain the relationship between stations, queues, and transporters and their pick-up and drop-off points. Pick-up and drop-off points are positioned strategically around the layout. A transporter travels between the pick-up and drop off points which have been associated with the transporter. A point may then act as entry or exit point to a station and/or queue if and only if that point is associated with the queue or station. An object may be associated with another object in SIMFACTORY using the domain editor.

To associate a station with a set of downstream queues, as is the case in the kitter station on the feeder line, the station option is selected in the editor and the cursor is placed over the kitting station. Then the downstream queues are selected in order of priority. Figure 16 illustrates the SIMFACTORY layout representing the kitchen appliance assembly line model. The station highlighted is the kitting station. The downstream queues which are associated with the station have a number over them, indicating the order in which a part should check for an available position in those queues.

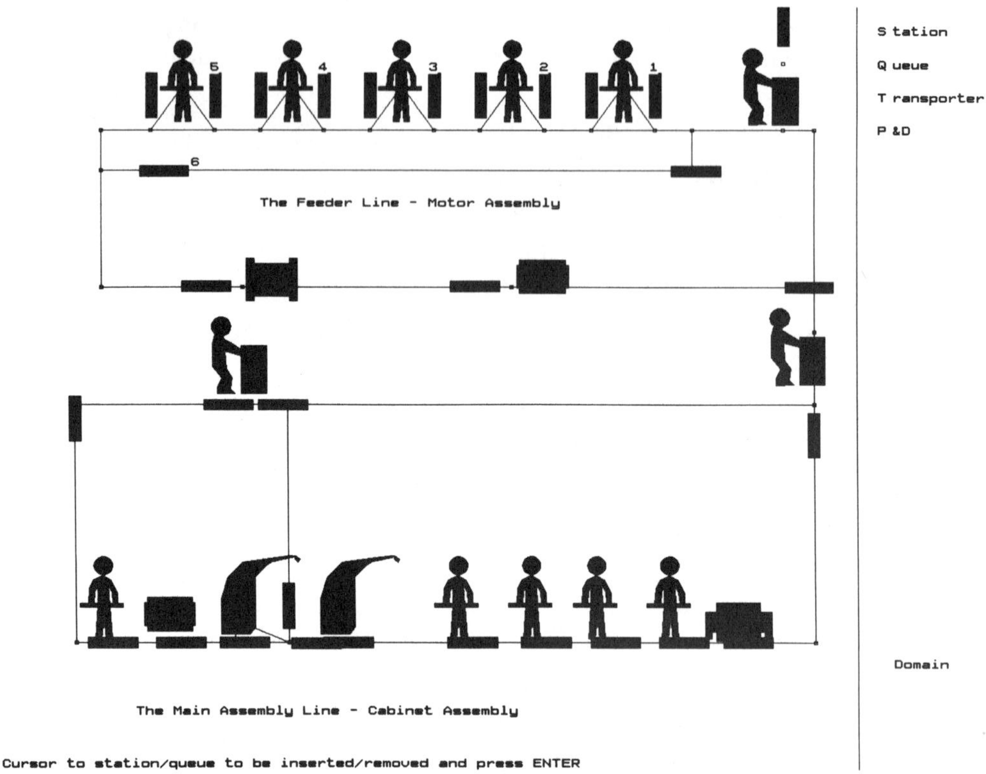

Figure 16. The graphical layout of the kitchen appliance assembly lines, developed within SIMFACTORY.

B. VERIFICATION OF THE SIMFACTORY MODEL

SIMFACTORY has the flexibility to fully model the kitchen appliance assembly lines with the exception of the transfer of parts to the five motor assembly stations. The decision as to which assembly station a part will travel is made at the kitting station. This had little effect on the overall model but had a minor effect on the utilization levels of the individual assembly stations. The processes of verification and validation of the model were combined because the modeler had a detailed and correct knowledge of the system. To verify the model, both the animation and simulation results were compared with the SIMAN/Cinema model. The final simulation results were very similar. The deviation can be explained by the problem of modeling the assembly stations, mentioned above.

TABLE 7
SIMFACTORY's Workstation Statistics, for the Feeder Line

SIMLAB—Assembly Line
1.47 FINAL PROCESSING STATION REPORT 12/06/88 10:55
(From 240.0 mins to 720.0 mins)

Station Name	Work%	Setup%	Interuptions% Prty	Interuptions% Passive	Idle%	Blkd%	Resrc%	Jobs
KIT-MOT	87.422	0.	0.	0.	12.578	0.	0.	
MOTOR-ASS-1	100.000	0.	0.	0.	0.	0.	0.	
MOTOR-ASS-2	100.000	0.	0.	0.	0.	0.	0.	
MOTOR-ASS-3	100.000	0.	0.	0.	0.	0.	0.	
MOTOR-ASS-4	99.025	0.	0.	0.	0.975	0.	0.	
MOTOR-ASS-5	62.894	0.	0.	0.	37.106	0.	0.	
RUN-IN	92.244	0.	0.	0.	7.756	0.	0.	
TEST-STATION	92.275	0.	0.	0.	4.351	3.374	0.	

⟨PgUp, PgDn, arrows : scroll⟩ ⟨Esc : exit⟩
SIMFACTORY r1.53

C. SIMULATION RESULTS FROM THE SIMFACTORY MODEL

In SIMFACTORY the user may specify elements of the system on which results should be collected and whether a data echo of the system description should be stored. SIMFACTORY displays a menu of eight system elements and allows the user to specify how many reports should be created. Each report is presented in tabular form and describes the performance of an element. SIMFACTORY can record statistics on resources, transporters, stations, throughput, raw materials, conveyors, and so on. Table 7 illustrates the station utilization statistics for the feeder line.

D. FEATURES OF THE SIMFACTORY PACKAGE

The main features of SIMFACTORY are its flexibility and ease of use. Its flexibility arises from the method it uses to describe a manufacturing system. It does not assume a standard station description where each station has an input queue and an output queue. This allows a higher degree of flexibility in part routing and was highlighted when modeling the general queue for the motor assembly stations in our kitchen appliance assembly line. Also, the use of the layout editor to define part flow simplifies and reduces the effort required to describe the work flow.

SIMFACTORY also allows a process stage to have more than one input and/or output. It is therefore possible to model both assembly and disassembly operations easily. Also, one process plan can feed into any number of other plans and hence model a complex series of part routes. Another features SIMFACTORY offers is error checking. When an error is found, an informative message is displayed on the screen. This can be helpful when trying to debug a model.

The animation of part movement and the simulation in general is slow. This may reflect the implementation of SIMFACTORY on IBM PC/XT.

VIII. PCModel

PCModel is a simulation package designed to model the physical movement of parts through a system and the times associated with the movements (White, 1988). Most simulation systems concentrate on the time delays associated with material flow, and any problems associated with collisions and the physical location of a part have to be explicitly programmed. PCModel, on the other hand, concentrates on the physical position of each part and statistics are collected by monitoring the number of parts which pass through a particular physical location and the time they spend there. Therefore, to monitor workstation utilizations a fixed point is associated with the station and the total time a part occupies that point is recorded.

The main components of PCModel are

- The source language preprocessor
- The graphical overlay editor
- The run-time interpreter

These three systems and the other supporting features are all accessed from PCModel's user interface. To simulate a system using PCModel, a model is built by describing the details of the process as a sequence of logical flow statements, called routings. To complement the routings, a graphical overlay is drawn which highlights the path a part may take and the position of each workstation.

PCModel has combined its animation and simulation capabilities into a single system. A part is represented by any character from the IBM PC character set and, therefore, part movement is restricted to the 80 × 24 character positions on the screen. To overcome this limitation, there is a windowing facility which increases the graphical size sixteenfold. To view a large model, a "window" may scroll over the complete model.

There are many other special features offered by PCModel, most of which have to be explicitly programmed. One very helpful option is the high degree of interaction allowed with the model during simulation. All time variables used by the model may be changed during the simulation run. Also, the reporting features complement this interaction by collecting hour-by-hour statistical reports so the change can be monitored, and statistics collected prior to the change are not affected.

The effectiveness of PCModel, as with all modeling systems, clearly depends on the accuracy of the model and the effort applied by the modeler. Therefore, prior to developing a model, the modeler should decide on the type of output he or she requires and the degree of interaction he or she requires during the simulation run.

A. PROCESS OF BUILDING THE MODEL USING PCMODEL

To develop a model using PCModel, an *application file* and a *graphical overlay* are needed. The application file contains all the information required to simulate the manufacturing system and the overlay is a graphical representation of the system. As PCModel models a system by describing the movement of a part from one physical location to another, the coordinates of each workstation and travel path have to be known. For the highest degree of accuracy the coordinates should be stated in units equivalent to the width of the part, as a part will always take up one unit of space. Therefore, all units used to describe the kitchen appliance model were converted into units equal to the size of the pallet on which the parts were transported. The same units are used in the graphical overlay and in the application file.

1. Creating a PCModel Application File

An application file describes a model using PCModel's own statements. There are two types of statements in the application file—directives and instructions. A *directive* is executed before the simulation begins and is used to assign system variables and parameters. An *instruction* is executed during simulation and is used to describe the routes parts take in the system.

There are two types of instructions—delaying and nondelaying. A *delaying instruction* holds an object for a specified time at that instruction, though the object may be moving on the screen. An example of a delaying instruction would be the *move* instruction. The *move* instruction is the most commonly used statement in the kitchen appliance assembly line model and it appears in two forms: absolute movement, where the part is sent directly to a position on the screen, at which point it waits for a period of time; and relative movement, where the part can be moved relatively in any direction, delaying for a specified period of

TABLE 8
Definition of Basic PCModel Instructions

MA(*LOC,TIME)	Move Absolute	Move to a location and delay a specific time. This statement is suitable for accessing a workstation and simulating the time delay at that station.
MU(@DIST,TIME)	Move Up	This statement moves an object up a distance relative to its original position and on its way it delays a specified time after each step up. There are similar statements for moving down, left and right: MD, ML, MR. Relative movements with periodic time delays are suitable for simulating material handling devices.
JP(:NEXT)	Jump	The Jump statement acts like a GOTO command in high-level programming languages. It unconditionally transfers the flow of control to another part of the program.
JB(*LOC,:NEXT)	Jump if Blocked	The Jump if Blocked statement is a conditional statement, which transfers the flow of control if the specified position(s) are blocked.
IF(A,GT,B,THEN,)	IF then	The If statement works in the very same way as it would in any computer language. Two symbols are compared using Boolean logic and, depending on the result, specified statements may or may not be carried out.

TABLE 9
Definition of Basic PCModel Directives

J=	Job description	The job directive specifies the manner in which jobs are released into the simulation, viz, part number, route number, the character to represent the part, conditions, and the number of parts to be introduced into the system.
L=	set wip Limit	In order to control the number of objects in the system, a work-in-progress limit is set.
O=	Overlay definition	The "O" directive specifies the graphical overlay file to be associated with the model.
U=	Utilization definition	The utilization directive defines the points that are to be monitored and the utilizations recorded.

time after each unit distance. Both types of *move* statements are illustrated in the program given below.

The *nondelaying* instructions are used to control the flow through a program. They can be conditional or nonconditional, and in some cases they delay a part until a certain condition is met. These instructions include both conditional and nonconditional *jump* statements as well as *if* type statements.

To illustrate how these statements work, a simple definition of a few statements and a small program are given below (see Tables 8 and 9). All of the variables inside the brackets are assigned appropriate values prior to use.

At this point it should be noted that, even though there is a *jump if blocked* statement, no two parts ever occupy the same position on the screen. If a part is scheduled to enter an occupied space, it will wait at its original position until the space becomes free. This facility proves invaluable when modeling queues and transporters.

All the abovementioned statements appear between two instructions *begin route* and *end route; br (route–number)* and *er*. To cope with multiple parts and

TABLE 10
PCModel Code Representing a Section of the Feeder Line Model

```
O= (=)                              ; Specifies the overlay file
L= (50)                             ; Specifies the work-in-progress limit
U= (*KITTER)                        ; Specifies that utilization of the position
                                    ;    represented by the symbol kitter
                                    ;    should be recorded
J= (01, #, 1, 1, 1, 0, 1250)        ; The directive states that job 1 will be
                                    ;    represented by # and follow route 1. The
                                    ;    job will repeat when job one is finished;
                                    ;    finally 1250 parts are to be created

        BR (1, *ENTRYLOC, %ARRIVAL)      ; A part will arrive at the entry location
                                         ;    every %arrival seconds
        (3, %HORIZ-CONV)                 ; Move left along conveyor
          MA (*KITTER, %KITTING)         ; Enter kitting station and wait
          ML (14, %HORIZ-CONV)           ; Move left along conveyor
: ASS#1                                  ; Entry point for assembly station one
          DN                             ; Do nothing
          JB (1, *ASSI#1, : ASSN#2)      ; Go to ASS#2 if ASS#1 buffer is full
          MU (2, %VERT-CONV)             ; Move up into the input buffer
        MA (*ASSEMBLY_1, %ASSEMBLY       ; Enter the assembly station and wait
•
•
•
          ER                             ;End route one
```

their different routes there are multiple route numbers. Each route is referenced by the directive job ("J"). Directives are used to set up the model and to specify the different symbols and parameters of the system. There were five directives used to model the kitchen appliance assembly lines and a brief description of each is given in Table 10.

There are many other directives and instructions but those outlined above should demonstrate the method of simulation used by PCModel. The model of the kitchen appliance assembly line uses the above commands and a section of the application file which describes the feeder line model is given in Table 9. In the program the symbols are used to specify times and locations. These are predefined at the beginning of the program.

2. Creating a Graphical Overlay within PCModel

The graphical overlay can be created or edited by the graphical editor. The editor can draw horizontal and vertical lines with either single- or double-line representation. All other graphics and text are taken from the IBM graphical character set. The overlay indicates the positions of workstations and highlights the path taken by an object. Any text required to describe the model is also

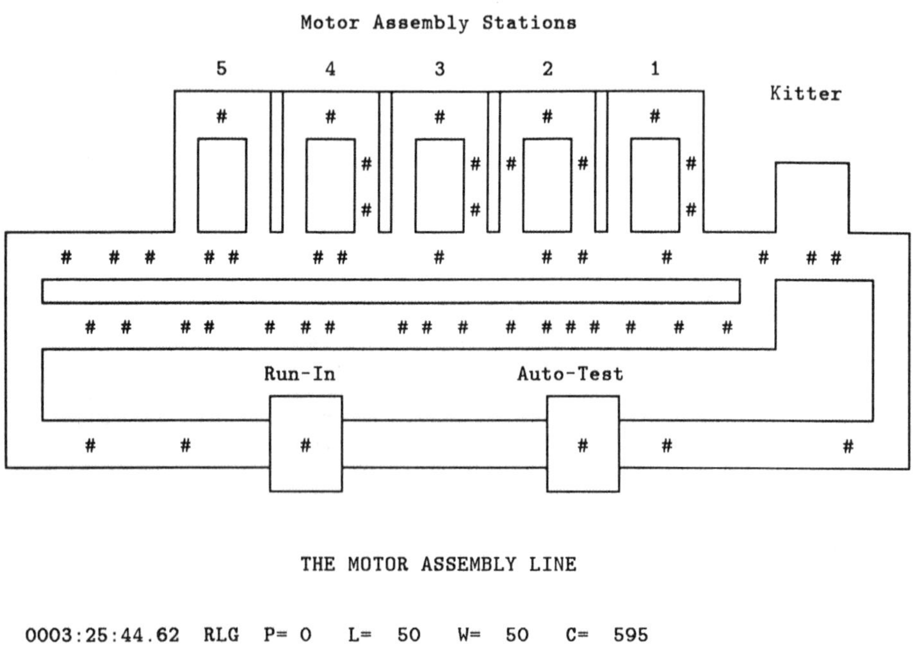

Figure 17. A graphical representation of the feeder line using PCModel.

added. Figure 17 illustrates the layout used to model the feeder line. Parts are represented by "#" and are dispersed throughout the line.

B. VERIFICATION OF THE MODEL DEVELOPED USING PCMODEL

Verification of the model was completed and the results were compared favorably with the results of SIMAN model. In developing the model it was possible to check its internal logic by analyzing the animation. The only problems encountered were associated with the positioning of buffers relative to workstations. It is very important when developing a model that all possible part routes are anticipated, as relative movement statements can cause a part to go off track unless they are checked thoroughly.

C. SIMULATION RESULTS FROM THE MODEL DEVELOPED USING PCMODEL

PCModel collects two types of statistics during simulation: work-in-progress information and resource-utilization statistics. Throughout the simulation the work-in-progress level, the maximum work-in-progress level, and the number of parts completed are displayed on the screen. The next set of statistics are the utilization statistics. The user specifies which locations are to be monitored. In the case of kitchen appliance assembly line model, the workstation and buffer statistics were monitored. Utilization statistics are collected on an hourly basis. Table 11 illustrated the utilization statistics for the feeder line. It can be seen that after the third hour the system has reached a steady state condition. At the same time the model has reached its work-in-progress limit and therefore only statistics after the third hour should be taken into account.

It is possible to explicitly program the collection of statistics and have them output to the screen or to a datafile. For example, the average size of a buffer could be calculated and displayed on the screen during simulation. Similarly, dynamic bar graphs can be programmed and hence drawn.

D. FEATURES OF THE PCMODEL PACKAGE

PCModel uses a different approach to simulation than SIMAN or SLAM II. It models a system by describing the part movement through physical locations. As a result, the animation and the simulation are combined and this has proved to be much faster to run than any other animation system reviewed in this paper. PCModel has another unique feature in that it allows a user to change variables during the simulation run. Thus it allows a higher degree of interaction during simulation than any of the other systems reviewed. On the other hand,

TABLE 11
Utilization Statistics Outputed by PCModel

Name	Hour 1	Hour 2	Hour 3	Hour 4	Hour 5	Hour 6	Hour 7	Hour 8	Hour 9	Hour 10
KITTER	99.42	90.76	87.48	87.52	87.44	87.52	87.44	87.42	87.44	87.52
ASSEM_1	98.37	100.00	100.00	100.00	100.00	100.00	100.00	100.00	100.00	100.00
ASSEM_2	96.58	100.00	100.00	100.00	100.00	100.00	100.00	100.00	100.00	100.00
ASSEM_3	94.34	100.00	100.00	100.00	100.00	100.00	100.00	100.00	100.00	100.00
ASSEM_4	86.58	100.00	100.00	100.00	100.00	100.00	100.00	100.00	100.00	100.00
ASSEM_5	90.35	100.00	77.30	64.30	64.24	64.24	64.31	64.17	64.15	64.33
RUN-IN	90.35	100.00	93.87	92.58	92.49	92.58	92.49	92.58	92.49	92.58
TEST	89.36	100.00	92.96	92.47	92.56	92.47	92.56	92.47	92.56	92.47
THRUPUT	148	167	168	167	168	167	168	167	168	167

the collection of statistics in PCModel is poor in comparison to other simulation systems. Only work-in-progress information and station utilizations are directly recorded. Any other statistics have to be explicitly calculated and saved.

IX. FALSIM

FALSIM (Flexible Assembly Line SIMulator) is a Petri net-based, data-driven general simulator (Wadhwa, 1988) which is written in FORTRAN and SLAM II. All input to the package is based on a Petri net description of the manufacturing system to be modeled. Before reviewing FALSIM we will give a short overview of the background to Petri nets.

Petri nets were designed specifically to model systems with interacting concurrent components, that is, the activities of one component may occur simultaneously with the activities of other components (Dubois and Stecke, 1983; Gawthrop, 1987). They were developed from the early work of Carl Petri, who formulated the basis for a theory of communication between asynchronous components of a computer system. He was particularly concerned with the description of causal relationships between events.

A Petri net is composed of the following four parts:

- A set of places P represented on a Petri net graph by circles
- A set of transitions T represented on a Petri net graph by straight lines
- An input function I
- An output function O

The input and output functions relate to transitions and places. Formally, a Petri net C is defined as the four-tuple $C = (P,T,I,O)$. As an illustration, consider the following example of a Petri net structure defined as a four-tuple and corresponding to the Petri net graphs shown in Figure 18:

$$C = (P,T,I,O)$$

$$P = \{p1,p2,p3,p4\}$$

$$T = \{t1,t2\}$$

$$I(t1) = \{p1,p4\} \qquad O(t1) = \{p2\}$$

$$I(t2) = \{p2\} \qquad O(t2) = \{p3,p4\}$$

The simple Petri net view of a system concentrates on two concepts, that is, transitions and conditions. *Transitions* are actions which take place in the system, the occurrence of these transitions being controlled by the state of the system. The state of the system may be described as a set of conditions. A

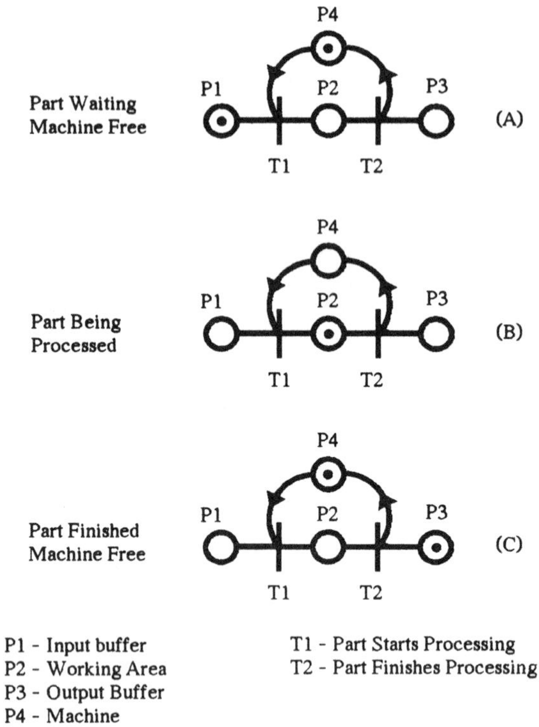

Figure 18. A Petri net representation of workstation.

condition is a logical description of the state of the system (Duggan and Browne, 1988).

For a transition to occur (or to fire), it may be necessary for certain conditions to hold, and these are termed the *preconditions* of the transition. The firing of the transition may cause other conditions, called *postconditions,* to become true. In a Petri net graph preconditions are modeled by tokens residing in an input place(s) of a transition and postconditions become true by placing a token in the output place(s) of a transition.

To demonstrate the use of Petri nets for modeling manufacturing and assembly systems, Figure 18 illustrates a Petri net representation of a manufacturing workstation and indicates three operating states:

• The first state shown in Figure 18(a) represents the conditions in the workstation prior to processing a part. The token in *p*1 represents a part waiting to be processed. The token in *p*4 represents an idle operator or machine which is available to work on the part. As both preconditions hold the transition, *t*1 may fire that is, the part can be operated on.

• The second state shown in Figure 18(b) represents the state of the workstation while a part is being operated on. The tokens representing the part and the operator have been removed from *p*1 and *p*4, respectively, and a token placed in *p*2 which represents the working area. This example uses timed Petri nets and therefore even though the precondition of *t*2 holds, the transition will not occur until a specified period of time elapses. This time represents the processing time for the part on the workstation.

• The third stage shown in Figure 18(c) represents the state of the workstation after a part has been processed. When the time period between *t*1 and *t*2 has elapsed, all the preconditions for *t*2 hold; therefore, the transition occurs and tokens are placed in the output places. The operator/machine is now available to work on a new part and the finished part proceeds to the output buffer *p*3.

When modeling manufacturing or assembly systems, it is difficult to model decisions using basic Petri nets. To remedy this problem, Wadhwa (1988) proposed decision point timed Petri nets (DPN), which are, as the name suggests, an extension to timed Petri nets. A point in a Petri net where a priority-based decision is to be made is referred to as a decision point and at this point one of three types of priority-based decisions can be made (see Wadhwa and Browne, 1989):

• *Token priority decisions* (DP1) select which token is taken from an input place (precondition) based on an attribute of the token.

• *Transition priority decisions* (DP2) select one of a number of possible transitions which can fire.

• *Place priority decisions* (DP3) select one of a number of possible output places (postconditions) to which to send a token.

Of the three decision types, only place priority decisions were used in the model of the kitchen appliance assembly line. To demonstrate how the decision Petri nets model this decision type, an example is given below.

Figure 19(a) illustrates a Petri net with one token in place *p*1. The preconditions for transition *t*1 to fire hold. After the firing of transition *t*1 the token

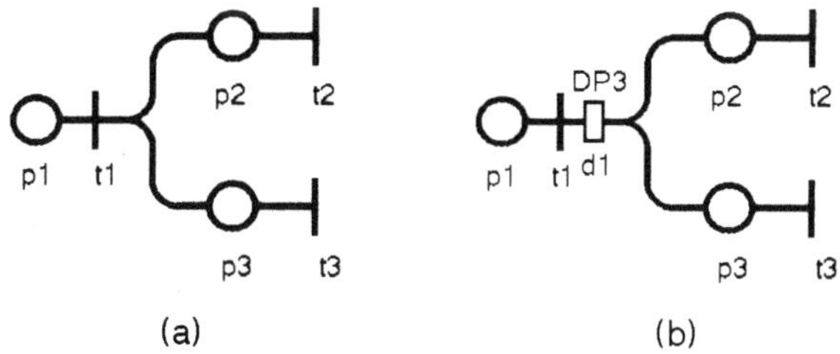

Figure 19. An illustration of the place priority decision point (DP3).

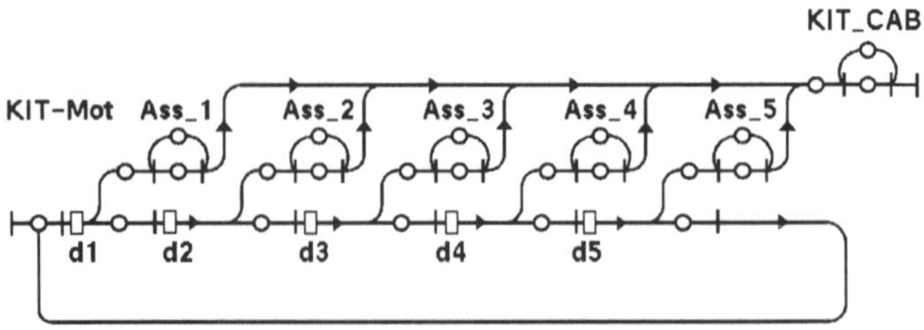

Figure 20. The DPN representation of the feeder line.

can either go to the output place *p*1 or to the place *p*2. Figure 19(b) illustrates a decision point extended timed Petri net for this case, with DP3 as the place priority decision point. It can determine probabilistically into which output place the token should be deposited. To model a nonprobabilistic decision, a FORTRAN subroutine which models the decision logic can be linked to FALSIM.

The Petri net graph representing the feeder line is shown in Figure 20 and illustrates the five decision points (*d*1–*d*5) on the line. The decision as to which place a token is sent is made at each decision point, that is, whether the part enters the input buffer of the nearest station or travels along the line to the next station. At the last decision point (*d*5), a part which fails to enter the input buffer is sent back to the start of the line and tries again.

In the model of the feeder line there is only one part type being processed. If, however, there were more than one part type it would be necessary to draw separate disjointed Petri nets to represent the routing for each part type. A disjointed Petri net models the flow of a single part type and, where two part types share common places, the common places use the same name. FALSIM input facilitates the use of disjointed Petri nets which simplifies the task of modeling manufacturing systems which process multiple part types.

FALSIM consists of two modules: an input module which prompts the user for the data to describe the proposed model, and the simulator which takes the system description and runs the simulation model. During simulation, the statistics are output to the screen and are continuously updated as the simulation run progresses.

A. PROCESS OF BUILDING THE MODEL USING FALSIM

Having developed a DPN graph of the model, the next step is to enter the data input requirements to FALSIM. This is a simple task as all data input can be read from the graph, except for the details of decisions made at each decision point. Where there is more than one part type flowing through the system, a separate DPN graph is drawn for each using the same places and transitions.

The data input to FALSIM can be considered under nine headings, of which the most important are outlined below.

• The Petri net logic table lists the input and output places for all transitions, the transitions that are associated with the output places, and the activity number associated with the transition. Activities represent the time delay between the firing of one transition and another transition. There is a table to correspond to the Petri net for each part type. The information in each table is taken from the appropriate DPN graph.

• The activity time table lists the times for all activities, such as process times and transportation times. Process times can be modeled by any SLAM II–supported statistical distribution.

• The decision point table lists the number of decision points and the decision rule used at each point. For a decision point the preceding transitions and the resulting places and transitions have to be entered. Also the parameters needed to make a decision are entered.

• The special transition/places table lists which places are associated with queues or stations and similarly which transitions represent process times as opposed to travel times.

• To initialize the system and make all machines available, tokens can be put in the appropriate places. By placing parts in queues (i.e., tokens in places) around the system, the problem caused by starting a simulation from an EMPTY state can be overcome.

• Finally, the Petri net boundness table lists the maximum number of tokens allowed in each place. In effect it limits the size of the queues in the system.

Having entered all the data taken from the DPNs, the model can be run and subsequently verified.

B. VERIFICATION OF THE FALSIM MODEL

To verify a model using by FALSIM, the actual Petri net has to be verified as the net defines all of the logic for the model. This task was relatively simple as the modeler had a good understanding of the real system. An interesting aspect of the model was the simulation of the five motor assembly stations on the feeder line. There were three possible DPN graphs. The first graph modeled the five workstations by putting five tokens in one place, to represent the five stations. This was not suitable, as statistics on the individual stations were required. The second suggested that the five stations should be modeled in parallel. Once again this did not correctly model the layout, as a part will travel along the conveyor passing the stations with full buffers and returning to its initial position if there is no room. The third model, which uses a cascading effect, was the most suitable option. The part travels along the net, attempting to enter the input place of one of the assembly stations. If it fails it continues along to the next decision point and tries again.

The model was verified by comparing the FALSIM model with the SIMAN model, which had been validated by the engineers who were working on the kitchen appliance assembly lines. The simulation results were quite similar, except during the warm-up period. The reason the two models differed during this period can be explained by the method used by FALSIM for modeling the conveyor system.

C. SIMULATION RESULTS OF THE FALSIM MODEL

Detailed simulation results are output by FALSIM to the screen during simulation. The statistics collected include

- Station/place utilization and blockage levels
- The average time per part in the system
- The average time between parts leaving the system
- The work-in-progress levels
- The number of parts produced

All statistics are updated at user-specified time intervals during the simulation run, and as a result the user can analyze the system in great detail. The user can also specify the maximum work-in-progress level at the beginning of

```
                    Simulation Running

                 Simulation Time 270 Minutes

Parts in System: 50              Maximum Parts in System: 50

                      Part Statistics

  Part Type    Number Finished      Mean TSYS      Mean Cycl Time
  1            791                  12.88          0.34


                  Resource Utilizations

Resource        Capacity              Utilization  Blockage
Kitter                                0.87         0
Assembly_1      1                     1.00         0
Assembly_2      1                     1.00         0
Assembly_3      1                     0.99         0
Assembly_4      1                     0.99         0
Assembly_5      1                     0.64         0
```

Figure 21. A sample screen outputed by FALSIM during simulation.

each simulation run. Figure 21 illustrates a FALSIM output generated during simulation of the feeder line.

D. FEATURES OF THE FALSIM PACKAGE

FALSIM has the capability to model any system that can be modeled using decision point extended Petri nets. As a result, it is likely to be more flexible than the general data-driven packages reviewed earlier. However, the cost of this flexibility is paid in terms of the fact that the user must have the background and training to develop Petri net models. At present FALSIM is a prototype system and can model only relatively small systems.

X. A Comparison of the Different Simulation Packages

To now we have reported on how a number of simulation packages and languages were used to model a kitchen appliance assembly line. It is clear that we have used two different types of simulator. SIMAN/Cinema and SLAM/TESS, on the one hand, are simulation *languages* and it is clear that the user of such languages needs to have a certain level of model building and simulation expertise. MAST, SIMFACTORY, and PCMODEL, on the other hand, are *data-driven modeling systems;* in effect they create a model of a particular manufacturing system based on a particular understanding of the dynamics of manufacturing and on the data supplied by the user. Whereas SIMAN/Cinema and SLAM/TESS are extremely flexible languages, they require a level of expertise and a commitment in terms of time to construct a model. The others are less flexible but allow a user with less simulation expertise to create a reasonable model more quickly.

In this section the differences between the packages, highlighted by the kitchen appliance assembly line model and the author's experience, are discussed. To highlight the comparison, selected tables are drawn up, displaying the rating given to each package under different criteria. The packages are compared under a number of headings and each main heading in turn has a number of specific subheadings. The main headings are as follows:

- The capability of the package to model the kitchen appliance assembly lines and other manufacturing and assembly systems
- The ease of use of the package
- The statistical output of the package
- The animation capabilities of the package
- The ability to link the package to other software products
- Any extra facilities the package might offer

A. BASIC CAPABILITIES OF EACH PACKAGE

It can be seen from Table 12 that all of the systems used, except MAST, are capable of modeling the assembly of parts, the disassembly of parts, and part failure. In the case of part and machine failure, each package supplies a wide range of statistical distributions to model resource and part failures, except for SIMFACTORY, which is rather weak in this respect. SIMFACTORY and FALSIM have a very small selection of scheduling rules. MAST, on the other hand, has a large selection and allows a user to code his or her own, using a FORTRAN subroutine. SLAM II also offers a selection of sequencing rules and again the user can program his own. MAST and SIMFACTORY have a wide range of options for modeling transport systems. They can model accumulating and nonaccumulating conveyers as well as conditional and nonconditional transfer of parts. The two simulation languages, SLAM II and SIMAN, can model extremely large systems when implemented on mainframe or minicomputers and in their PC implementation can model relatively large systems. SIMFACTORY can simulate a system without storing the complete model in primary memory and therefore can model large systems. The size of a model developed for PCModel depends on the CPU used but in general it can model relatively large systems. MAST is relatively slow and SIMFACTORY is very slow, even with the animation facility turned off.

TABLE 12
A Comparison of Modeling Capability

Ability to model	SIMAN	SLAM II	MAST	SIMFACTORY	PCModel	FALSIM
Assembly	Yes	Yes	No	Yes	Yes	Yes
Disassembly	Yes	Yes	No	Yes	Yes	Yes
Part failure	Yes	Yes	No	Yes	Yes	Yes
Machine failure	Yes	Yes	Yes	Yes	Yes	Yes
Scheduling rules	Program	Very good	Good	Fair	Program	Fair
Transport						
• AGV	Very good	Program	Very good	Very good	Program	Good
• Conveyers	Very good	Program	Very good	Very good	Program	Good
Nonaccumulating			Yes	Yes		
Accumulating			Yes	Yes	Yes	
• Conditional	Program	Yes	Yes	Yes	Program	Yes
• Nonconditional	Yes	Yes	Yes	Yes	Yes	Yes
Model limits	Very large	Very large	Medium	Large	Large	Small
Run time	Very fast	Very fast	Fair	Very slow	Fast	Very fast

B. COMPARISON OF THE EASE OF USE OF THE PACKAGES

In our view, of the packages studied, SIMFACTORY is the most flexible and probably the easiest to use, although it lacks a selection of control strategies. MAST, on the other hand, has a wide variety of control strategies and allows a user to define his own if appropriate, but it is not as flexible as SIMFACTORY. SIMAN and SLAM II are similar. SIMAN has more commands but may require more code to model a similar model. SIMAN and SLAM II do not have menu-driven input but they have supporting software for generating code. FALSIM is different from the other packages in that the user has to be able to develop a model using DPN, which increases its flexibility but also increases the effort needed to develop a model. PCModel has a user interface which offers many options, but the model is developed within a text editor. The PCModel language commands are relatively easy to use but great care is needed to ensure that the physical coordinates are correct. Table 13 is a summary of our view of the relative "ease of use" of the systems we have reported. We have defined "ease of use" in terms of whether the system is menu-driven, whether on-line help is available, the level of programming skills required to use the system, the difficulty of model building, the flexibility of the system, how decision rules are incorporated, and the ease with which the model may be debugged. It should be noted that, in all tables, the word "Program" is used when a problem may be programmed but is not explicitly defined by the system.

TABLE 13
A Comparison of the Ease of Use of the Packages

User-friendly	SIMAN	Cinema	SLAM II	MAST	SIMFACTORY	PCModel	FALSIM
Menu-driven	None	Excellent	None	Good	Very good	Limited	Good
On-line help	None	Very good	None	None	On version 1.6	Limited	None
Program skills	Medium	None	Medium	None[a]	None	Medium	None
Model building	Hard	Simple	Hard	Simple	Very simple	Hard	Medium
Flexibility	Very high	High	Very high	Limited	Flexible	High	High
Decision rules	Program		Program	Good	Fair	Program	Fair
Ease of debug	Hard		Hard	Fair	Easy	Hard	Easy

[a]MAST generally requires no programming skills but the user may link a FORTRAN subroutine to MAST to model a decision rule.

TABLE 14
A Comparison of Output Statistics

Output statistics	SIMAN/Cinema	SLAM II	MAST	SIMFACTORY	PCModel	FALSIM
During simulation	Cinema/Yes	Yes	No	Limited	Yes	Yes
After simulation	Yes	Yes	Yes	Yes	Yes	Yes
User-defined	Yes	Yes	No	Partially	Yes	Yes
Fixed	No	Yes	Yes	Yes	No	Yes
Graphical	Cinema	No	Yes	Yes	Program	No
Detail	Very high	Very high	High	High	Fair	High
Blocking reports	Program	Yes	Yes	Yes	Program	Yes

C. STATISTICAL OUTPUT FROM THE VARIOUS SYSTEMS

It can be seen from Table 14 that all of the simulation packages have very good report facilities except for PCModel, where only station utilizations are recorded. Any specific statistics and output have to be calculated during simulation. FALSIM automatically generates statistics during the simulation run and a user may specify those statistics to be recorded. In SIMAN the user has to specify the statistics to be collected and must explicitly program the collection of blocking statistics. SLAM II, on the other hand, automatically records most statistics, although the user may specify some others, if desired. MAST and Cinema automatically produce graphical output during animation and in Cinema, dynamic graphs can be created. In PCModel, simple bar charts and user-defined statistics can be output during animation.

D. ANIMATION CAPABILITIES

In recent years, with the great developments in computer graphics and the availability of relatively cheap microcomputers, interest in animation has greatly increased and now many suppliers of simulation packages and languages offer animation facilities with their products. In our view, animation capabilities are important additions to the "tool box" of the model builder from two points of view:

• Animation is a useful activity in terms of convincing nonsimulation people of the inherent value of a simulation model. It allows the model builder to display in some sense the workings of his or her model and in this way promotes the credibility of the overall approach.

• Animation is useful to the model builder in terms of checking the internal logic of the model, that is, in the verification stage of the overall validation procedure.

TABLE 15
A Comparison of Animation Features

Animation	Cinema	MAST	SIMFACTORY	PCModel
Real time	Yes	No	Yes	Yes
Post processor	No	Yes	No	No
Ease of construction	Very easy	Fair	Easy	Program
Icons	Very good	Limited	Good	None
Panning			Yes	Yes
Zooming			Yes	
Skip ahead	Yes	No	Yes	No
Speed	Fast	Slow	Very slow	Very fast
Detail	Very good	Medium	Good	Fair
Statistical output	Very good	Graphical	Graphical	Fair

Of all the packages we have studied, Cinema offers the most detailed and flexible animation system. It is relatively fast and the user can create any level of detail he or she requires. For animation speed, PCModel is very good but its part icons are characters and therefore not realistic in appearance. PCModel code defines the animation and therefore the construction effort is shared with the effort required to create a model. The user interface to Cinema and SIMFACTORY is very good but the SIMFACTORY icon editor is relatively crude. MAST is the only system analyzed which animates the events of the simulation after the simulation is finished, therefore allowing the animation to be examined more than once without rerunning the model (see Table 15).

E. SUPPORT FOR EXTERNAL INTERACTION

Many manufacturing systems have particular complexities which may not be possible with the core facilities in the various packages. In cases such as this

TABLE 16
A Comparison of Access to External Packages

	SIMAN[a]	SLAM II[a]	MAST	SIMFACTORY	PCModel	FALSIM
High-level language	FORTRAN	FORTRAN	FORTRAN	None	None	None
Database	Yes	Yes	No	No	No	No
Datafiles	Yes	Yes	No	No	Yes	No
Other packages	Yes	Yes	No	No	Lotus	No

[a]SIMAN and SLAM II can access any FORTRAN compatible package, database, or datefiles.

it is important that the user have access to external facilities to create code to model the particular complexity in question.

SLAM II and SIMAN facilitate direct access to FORTRAN subroutines and therefore have access to any package that can be called or connected with FORTRAN. Similarly, either language can access databases or datafiles through FORTRAN. PCModel can use its own functions to access datafiles or Lotus 123 files. MAST can also access FORTRAN to a limited degree. A user can code decision rules in FORTRAN and link them to the MAST model (see Table 16). PCModel/GAF, which is a high-resolution graphics version of PCModel, can use CAD drawings developed by certain computer-aided design packages, including AutoCAD and AutoSketch.

F. AVAILABILITY OF SUPPORT FACILITIES

Table 17 lists the supporting software available for the various packages and languages we have used in our case study. SIMAN has four supporting systems which were referred to earlier and are listed again in Table 17. SLAM II is supported by TESS. MAST is in a unique position because the data going into it and the statistics coming out from it are analyzed. MAST carries out a static capacity check on the system description before simulating the model and suggests possible changes when inadequacies occur. Similarly, it has an expert system which analyzes the simulation results and suggests some changes. However, the expert system is limited in the range of problem areas which it looks at and in our experience is not very specific in the suggestions it offers.

TABLE 17
A Comparison of Extra Facilities

Extra facilities	SIMAN	SLAM II	MAST	SIMFACTORY	PCModel	FALSIM
Supporting software	Cinema BLOCKS ELEMENTS OUTPT	TESS			Lotus PCModel/GAF	
Input analysis			Static capacity check			
Output analysis			Limited expert system			

XI. Conclusion

In this paper we have looked at the role of simulation modeling as a tool to support the detailed design of flexible manufacturing and assembly systems. Specifically, we have taken a relatively simple case study of an assembly line and shown how a simulation model was used to help to "fine tune" the design of that line. Further, we have reviewed a range of simulation models and packages, the majority of which are commercially available, and indicated their usefulness in the context of our case study.

It is clear to us that simulation is an important tool for the manufacturing systems designer. Further, the availability of sophisticated simulation systems, many of which were designed specifically with the needs of manufacturing engineers in mind, has made simulation accessible to all manufacturing engineers and systems designers. Data-driven simulation models, together with existing capabilities in terms of animation and model debugging, as well as emerging facilities in terms of the analysis of model output, have made simulation a widely useful and usable tool.

References

Anonymous (1984). "Cinema User Manual." System Modeling Corporation, State College, Pennsylvania.

Anonymous (1986a). "BEAM User Manual." CMS Research Inc., Brooklyn Center, Oshkosh, Wisconsin.

Anonymous (1986b). "MAST User Manual." CMS Research Inc., Brooklyn Center, Oshkosh, Wisconsin.

Anonymous (1986c). "SPAR User Manual." CMS Research Inc., Brooklyn Center, Oshkosh, Wisconsin.

Anonymous (1987a). *Prod. Eng.* December, pp. 30–32.

Anonymous (1987b). "SIMFACTORY with Animation; User's and Reference Manual." CACI, Inc., Los Angeles, California.

Anonymous (1988). *Simulation* **51,** 136–156.

Browne, J., and Davies, B. J. (1984). *Int. J. Prod. Res.* **22,** 335–357.

Browne, J., and Rathmill, K. (1989). *Proc. Int. Conf. Simul. Manu., 4th.*

Browne, J., Chan, W. W., and Rathmill, K. (1985). *Ann. Oper. Res.* **3,** 207–237.

Dubois, D. N., and Stecke, K. E. (1983). *Proc. IEEE Conf. Decision Control,* 22nd vol 3.

Duggan, J., and Browne, J. (1988). *Inst. Electr. Eng.* **135,** 239–247.

Gawthrop, P. J. (1987). *Int. J. Adapt. Control & Signal Process.* **1.**

Hall, D. N., and Stecke, K. E. (1986). *In* "Flexible Manufacturing Systems, Operations Research Models and Applications" (K. E. Stecke and F. Suri, eds.), pp. 145–156. Elsevier, Amsterdam.

Pegden, D. C. (1984). "Introduction to SIMAN." System Modelling Corporation, State College, Pennsylvania.

Pritsker, A. B. (1984). "An Introduction to Simulation and SLAM II," 2nd ed. System Publishing Corporation, West Lafayette, Indiana.

Schlesinger, P. (1974). *Proc. Summer Comput. Simul. Conf. 1974,* vol. 1, p. 927.

Standridge, C. R. (1984). "TESS User's Manual." Pritsker and Associates, Inc., West Lafayette, Indiana.

Stecke, K. E. (1983). *Manage. Sci.* **29** 273.

Stecke, K. E., Dubois, D., Browne, J., Sethi, S., and Rathmill, K. (1984). *FMS Mag.* **2,** 114–117.

Suranjan, D., Nof, S., and Whinston, A. (1985). *In* "Decision Support Systems 1," pp. 37–56. Elsevier/North-Holland, Amsterdam.

Wadhwa, S. (1988). Development of generalized simulators for design evaluation of robot based flexible assembly systems. Ph.D. Thesis, Department of Industrial Engineering, University College, Galway, Ireland.

Wadhwa, S., and Browne, J. (1989). *FMS Mag.*

White, D.A . (1988). "PCModel—Personal Computer Graphic Modeling Systems—User's Guide" (Version 8). Simulation Software Systems, San Jose, California.

CHAPTER 9

Flexible Manufacturing Systems: A Survey of the U.S. Applications

Ajit K. Patankar
Allied-Signal Aerospace Company
Phoenix, Arizona

I. Introduction

The United States is considered as one of the pioneers in the flexible manufacturing systems (FMS) area. The U.S. leadership in FMS technology can be viewed as a natural extension of the U.S. developments in computers, communications, and numerical control technology. The concept of flexible manufacturing was developed in late 1960s and early 1970s. Some of the earliest users of FMS in the United States were Allis Chalmers and Rockwell Corporation (Klahorst, 1984). Along with the changes in FMS technology, the FMS management philosophy has also significantly changed in the last 20 years. Hence, this paper reviews technological as well as management aspects related to the U.S. installation of FMS.

The U.S. FMS builders or suppliers and the user base are diverse. Usually, the suppliers fall in one of the following categories:

1. Machine tool companies diversifying in FMS business, for example, Kearney and Trecker and Cincinnati Milacron. The NC machine tool technology has advanced significantly and these machine tools include many features which facilitate their integration in FMS. Some of these features are high-level communication and software interface to NC controllers, program storage capacity, tool storage and automatic tool change capability, error detection and correction ability, and so on.
2. Integration or automation software vendors, for example, Westinghouse Automation, Litton Industrial Automation, CIMCORP, and ITP Boston. Some of these companies diversified from their base in the NC controller market while other companies, like ITP Boston, were specifically started to develop and market automation systems.
3. Large corporations internally developing FMS, for example, the General Electric plant in Lynn, Massachusetts. The usual strategy is to buy standardized NC machine tools and computer hardware and then develop an FMS control system or software in-house using a high-level programming language such as "C." As FMS control systems essentially represent internal business policies and procedures, this approach avoids modifying business policies to suit external software design. Also, as mentioned above, advances in machine tool technology have also helped in promoting in-house development of FMS technology.

The U.S. strength in the FMS technology is a result of both private and government initiatives. In the private sector, companies like Vought (LTV) and General Dynamics pioneered significant aspects of the technology. The U.S. Air Force, through its Aeronautical and Materials Laboratories at Wright Patterson base, has provided considerable financial and technical incentives (Parker, 1987). Generally, aerospace companies—including airframe, aircraft engine, and aerospace component manufacturing companies—have been leaders in FMS installations. However, at present major installations exist in automotive, agriculture equipment, machine tool manufacturing, and electrical equipment manufacturing areas. Nonaerospace manufacturing is expected to be a major growth area for FMS installations in the coming years (Palframan, 1987).

It is important to consider the difference between flexible manufacturing systems and flexible manufacturing cells (FMC). The FMC are in effect FMSs with less automation; they are also simpler and cheaper to install. But the underlying principle is the same: flexible production. As FMS installation involves significant investment in training and equipment, many companies are preferring to first install FMC to get some learning experience with simpler cells rather than jumping immediately to FMS.

It is difficult to select any one FMS as a typical U.S. installation. The reason is not only the large number of FMSs installed over the last several years, but also the evolution of the concept of FMS itself over these years. The FMS selected in this paper will be surveyed regarding the following important factors: physical architecture, control system, operational aspects, and significant technological advances.

II. Flexible Manufacturing Systems Installations in the United States

It is difficult to define a U.S. FMS installation due to the international nature of the FMS industry. It is the intent of this paper to explain some of the significant FMS developments in the United States. Therefore, a simplistic U.S. geographical location will not be a valid selection criterion. In this report, the following guidelines are considered in selection of the FMS installations based in the United States:

- The user and FMS developer are both based in the United States. For example, the FMS at the Caterpillar plant in Aurora, Illinois, is supplied by Cincinnati Milacron (Dutton, 1989).
- The user is based in the United States but the FMS developer is based outside the United States. For example, aircraft components FMS at the Boeing plant in Auburn, Washington, is supplied by Shin Nippon Koki of Japan (Palframan, 1987).
- The user and FMS developer are both based outside the United States, but FMS is used in the U.S.-based factory of the parent corporation. For example, FMS for the Mazak Machine Tool plant in Florence, Kentucky, is also supplied by Mazak (Palframan, 1987).

The last category has been excluded in this survey as it reflects development of FMS technology outside the United States. There are some U.S.-based FMS suppliers who are owned by companies outside the United States. For example, CIMCORP Inc. of Aurora, Illinois, is owned by a parent company based in Finland (information based on personal communication in 1988). The FMSs supplied by these companies are considered to be U.S.-based if all essential technical development was done within the country.

TABLE 1
List of FMS Users in the United States[a]

Company	FMS location	Product	Main supplier
		Aerospace: Engines	
Avco Lycoming	Stratford, CT	Engine components	Kearney & Trecker
Pratt & Whitney	Columbus, GA	Engine components	
General Electric	Wilmington, NC Lynn, MA	Engine components	Lodge & Shipley
		Aerospace	
Boeing	Auburn, WA	Aircraft components	Shin Nippon Koki
FMC Corporation	Brea, CA Aiken, SC	Joints and Hoses components for military vehicles and rocket launch systems	Cincinnati Milacron
General Dynamics	Fort Worth, TX	Components for F-16 aircraft	Westinghouse
Lockheed Georgia	Marietta, GA	Airframe parts	White-Consolidated
Vought	Dallas, TX	Aircraft fuselage for B-1 bomber	Cincinnati Milacron
McDonnell Douglas	Torrence, CA St.Louis, MO	Aircraft components Missile bodies	Cincinnati Milacron Giddings & Lewis
Westinghouse Electrical Systems	Lima, OH	Aircraft generator components	Westinghouse
		Equipment	
Caterpillar	East Peoria, IL Decatur, IL	Tractor components	Cincinnati Milacron Giddings & Lewis
John Deere	Waterloo, IA	Tractor parts	Kearney & Trecker
Detroit Diesel	Detroit, MI	Piston engines	Lamb Technicon
Ford	New Holland, PA	Sheet metal parts for farm equipment	Trumpf
Ingersoll Rand	Roanoke, VA	Hoist equipment	White-Sunstrand
Mack-Truck	Hagerstown, MD	Transmission Housing and casing	Kearney & Trecker
		Machine tool	
Cincinnati Milacron	Mt. Orab, OH	Parts for plastics Processing equipment	Cincinnati Milacron
Mazak	Florence, KY	Parts for machine tools	Mazak
White-Sunstrand	Belvidere, IL	Machine tool parts	White Consolidated

TABLE 1 (continued)

Company	FMS location	Product	Main supplier
		Other industries	
Allen-Bradley	Milwaukee, WI	Motor starters	Kearney & Trecker
Borg Warner	York, PA	Compressor components	Comau
Federal Mogul	Littiz, PA	Bearings	
G.E. Medical Systems	Milwaukee, WI	Parts for medical equipment	Trumpf
Mercury Marine	Fond Du Lac, WI	Outboard motor crankcase and block	Kearney & Trecker
NY Air Brake	Watertown, NY	Brake systems	
OMC-Evinrude	Milwaukee, WI	Outboard motor parts	Swedish Machine Group
Onan	Minneapolis, MN	Electrical switchgear for generators	Trumpf
Remington Arms	Ilion, NY	Firearm parts	
Union Special	Huntley, IL	Sewing machine bodies	Kearney & Trecker
Westinghouse Electric Corp.	Pittsburgh, PA	Wire harness for electrical equipment	
Xerox	Webster, NY	Panels for copiers	

[a]Source: based on Palframan (1987), Dutton, (1989), Runewicz (1987), Sturges (1987), and Venkatesh (1990).

A. FLEXIBLE MANUFACTURING SYSTEM SELECTION CRITERIA

Table 1 lists major FMS installations in the United States. This table supports many of the observations made earlier about FMS installations in the United States. The survey shows that the U.S. FMS installations range from aerospace applications at Vought or General Dynamics to office automation applications at Xerox Corporation. Considering such a diversity and depth of applications, the case studies in this paper have been selected based on the following factors: type of industry served, uniqueness of application, and technological advances.

Usually, an FMS is associated with discrete part machining operations. But in the last 10 years, several significant applications have been developed in other areas, such as sheet metal cutting, welding, and assembly operations. Even automated inspection systems can be considered to be FMS as inspection is an important part of manufacturing operations. Typical U.S. installations for non-machining applications will be also reviewed.

III. Flexible Manufacturing Systems for Aerospace Machining Application

A. CATEGORY

The FMS at the General Dynamics (GD) facility in Fort Worth, Texas, has been selected in this category. This FMS is used for machining of discrete parts (Parker, 1987).

B. INTRODUCTION

This FMS was developed as a result of the combined efforts of GD and the Air Force Wright Aeronautical Laboratories/Materials Laboratory (AFWAL/ML). As this FMS is a part of a overall computer integrated manufacturing (CIM) architecture or strategy, it cannot be labeled as an island of automation. Also, this FMS is important because it is both a commercial and a research and development venture. One of the goals of this FMS was to advance the American machine tool and factory automation industries. Westinghouse Automation Division was the main supplier.

C. PHYSICAL CONFIGURATION

DeVlieg Machine Co. supplied the machine tools for this FMS, which is comprised of six, five-axis DeVlieg 430R machining centers. The pallet changers at the front of each machine automatically perform pallet exchanges between the AGV, pallet queueing stations, and the machine bed. A second pallet changer at the rear of the machine tool is used to exchange tool carousel pallets with the AGV.

Part dimensions are inspected on LK Tool horizontal-arm coordinate measuring machines (CMM), which are fully integrated with the FMS to provide for unattended operations. Cutting tools are stored in a central tool crib. Required tools are manually loaded on a tool pallet carousel. When these tools are loaded, a pallet changer in the tool crib automatically loads the tool pallet onto an AGV for delivery to the appropriate machine. At the machine, a GMF robot loads these tools into the tool magazine and unloads dull or unneeded tools to be returned to the tool crib.

The entire basic FMS is controlled by a cell control system based on artificial intelligence (AI) and distributed processing technology. Automated scheduling and control has been achieved. The FMS control system is linked with plant-level control systems to achieve integration of FMS operations in the overall factory.

D. OPERATIONAL DESCRIPTION

Parts are loaded to fixtures at one of two load/unload stands. Fixtures are permanently mounted to tooling cube tombstones, which in turn are mounted on part pallets. The fixture/cube/pallet assemblies are stored in a pallet storage area. This assembly is retrieved by the tracked automatic pallet changer (APC) and delivered to one of the two load/unload stands. At this place it is picked up by an AGV.

E. MACHINING CENTERS

The DeVlieg machines used in this FMS are unique as they have several built-in features to facilitate unmanned or unattended operations. These CNC machines have five-axis capability and are equipped with an automatic tool changer.

The tool magazine has a capacity of 114 tools. The tool magazine was designed to be compatible with the tool interchange robot. In conventional systems, tool identity is verified manually at this stage. To avoid this manual operation, an elaborate tool identification system is included. Each FMS tool holder is equipped with a EEPROM memory chip to store tool identity. The robot presents the tool to a reader before placing the tool in the magazine. A second reader head verifies tool identity again before the tool is actually changed into the spindle. This prevents any tool rearrangement errors caused while the machine is down or in manual mode.

Self-monitoring and checking capabilities are included in the machine tools. Each machine has a touch probe that can be used to automatically verify part pallet identity by touching a series of bolts that represent the pallet identification number in binary code. The probe also is used to verify pallet and fixture alignment as well as some in-process inspection operations. A second probe, permanently mounted on the machine bed, is used to verify tool length and detect broken tools.

Machine tools are controlled through an Automation Intelligence Producer II CNC. This CNC has several features designed for unmanned operation in FMS. Each CNC has a hard disk for storage of part programs, which can be used if the communication line fails. The CNC monitors tool wear by keeping track of cutter use. Spindle horsepower measurement is used to detect unexpected cutter wear. The actual and commanded positions of a machine axis are also tracked. If the difference exceeds the preset limit, an alarm is sent to the FMS control system. This prevents generation of scrap parts in an unmanned mode.

F. INTEGRATED INSPECTION EQUIPMENT

This FMS uses CMM manufactured by LK Tools. Inspection is considered an important part of FMS operations and thus is fully integrated in FMS. The CMM has several special features to achieve integration.

The FMS parts are machined on a horizontal spindle machining center and remain on the same fixture for inspection. To utilize the same horizontal fixture, the CMM is designed with a horizontal arm. This is a deviation from conventional practice of vertically oriented, gantry robot-type CMMs. The standard CMM rotary table is modified so that the DeVlieg pallet changer can automatically perform load and unload operations. Additional flexibility has been achieved by using an automatic probe changer. This is necessary as a large variety of FMS parts cannot be inspected using only a single stylus probe or sensor. Conventional touch type and the recent laser probes are both used in the system.

Inspection data collected by CMM are transferred to the FMS controller. At the FMS controller, statistical analysis is performed. The data are subsequently passed to plant-level quality assurance systems.

G. MATERIAL HANDLING SYSTEM

The system is comprised of two battery-operated AGVs and pallet changers. The AGV was supplied by Control Engineering, a subsidiary of Jervis B. Webb Company. The pallet changers were supplied by DeVlieg, the supplier of machine tools.

The AGVs can automatically charge batteries through contact plates mounted on the floor at two system locations; as batteries can be charged any time the vehicles are idle. This method is known as "opportunity" charging. This has enabled the FMS to operate with only two AGVs. Standard DeVlieg pallet receiving tables are installed on top of the table lifting mechanism of the AGV, which is completely passive during the pallet changing operation. A pallet changer does all the work of sliding a pallet onto the AGV or removing the pallet from it.

The AGVs are used to transport both part pallets and tool pallets in the FMS. The pallet with tools or parts appears exactly the same to the AGV. Fixtures are permanently set up on pallets. They are changed only for repair or modifications. Although this may have resulted in higher fixture costs, manual intervention necessary to setup fixtures on pallets has been avoided. But to use this arrangement, an automated pallet storage and retrieval mechanism is necessary.

Conventional fixture design is intended to satisfy machining requirements. However, in this FMS, fixtures have been designed to facilitate automated robot loading. The robot uses pins on the fixtures to place raw material billets on them. By a simple tilt of the fixture-holding table, the billet weight is directed toward the locating pin. This avoids the necessity of special billet preparation or holding mechanisms to keep the billet in place.

H. CONTROL SYSTEM

The control system essentially determines the performance level of the complete FMS. This control system also integrates FMS operation within the plant. The system has been designed to optimize the performance of the FMS and at the same time achieve integration with the rest of the plant. The system architecture is shown in Figure 1. The control system software and hardware are described in the following sections.

I. HARDWARE

This system uses a combination of proprietary and off-the-shelf hardware. The VAX/VMS system is used for the main cell control functions. The control system functionality is distributed, permitting use of a dedicated VAX computer

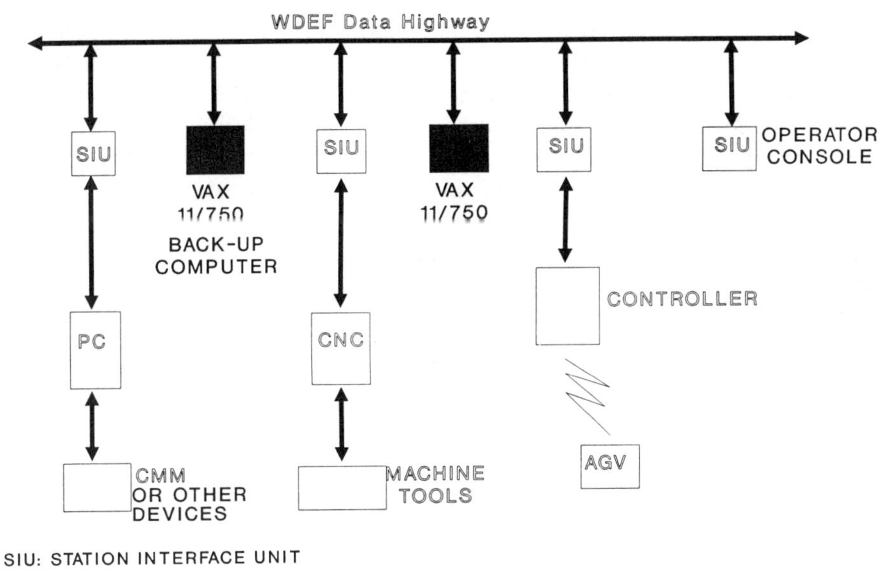

Figure 1. Flexible manufacturing system for aerospace application—control system architecture.

for FMS schedule generation. The second VAX computer maintains the FMS database and communicates with the plant-level shop floor control system. A third VAX serves as a backup for the other two VAX computers.

The VAX computers described above do not perform any real-time functions. The real-time functions are performed by processors called station interface units (SIU). The SIU hardware is based on Intel Multibus architecture with added functionality provided by proprietary hardware from Westinghouse. The SIUs are installed at machine tools, CMMs, AGVs, tool cribs, and at the load/unload station. Thus SIU is a key to the integration of dissimilar hardware and software. The SIUs communicate with devices under their control and ensure real-time execution of the schedule. SIUs also collect data from devices. These data are transmitted to processors called man–machine interfaces (MMI). These MMIs essentially act as engineer or operator consoles. Similar to the SIUs, MMIs are also based on Intel Multibus architecture; however, MMIs are equipped with a color monitor and a keyboard.

The communication network is based on a Westinghouse distributed processing family (WDPF) protocol called a data highway. This is a 2 MBPS, token bus type of network. The VAX computers are linked to the network through a direct Unibus connection. All other processors, such as SIUs and MMIs, are connected directly to this network.

J. SOFTWARE

Software is designed to utilize the distributed hardware architecture described above. The two main components of the software, the scheduling and status monitoring systems, will be described here in detail.

The FMS control system is designed for automatic schedule generation based on input from a plant-level control system. A rule-based expert system is used for FMS scheduling. The scheduler was developed using automated reasoning tool (ART) software from Inference Corporation. This software runs on a dedicated VAX computer. The scheduler first attempts to meet the due dates set at plant level. After due date objectives are achieved, the schedule attempts to minimize secondary objectives such as maximum machine use, minimum bottlenecks, or reduction of unnecessary AGV moves. The plant control system daily downloads a list of orders to the FMS control system. The schedule is primarily prepared once a day; however, the scheduler automatically changes the schedule based on the dynamic environment.

As described in the hardware section, SIUs are used to collect status information about various devices. Data collected by SIUs are used by MMIs to provide status information to operators and engineers. Software on MMIs converts status information to a graphical display format. As the status is updated once every second, the net effect is almost like an animated display of the system. Status monitoring can be performed without VAX involvement. Unlike FMS control systems, status monitoring is completely separated from the real-time control function.

K. SUMMARY

This is one of the most advanced FMSs in the United States. Completely unmanned operation has been achieved by several enhancements in the current technology. Some of the important enhancements are

- A scheduling system based on artificial intelligence and closely coupled with the plant-level control systems
- Robotic loading and unloading of parts to fixtures
- Automated storage and retrieval of raw material, finished parts, and tooling

The design for integration of FMS with the complete factory indicates an important milestone in the technology development.

IV. Flexible Manufacturing Systems for Sheet Metal Fabrication Application

A. CATEGORY

This FMS at the General Electric facility in Salisbury, North Carolina, is used for fabricating sheet metal enclosures and associated components for electrical equipment (Runewicz, 1987).

B. INTRODUCTION

This is one of the pioneering applications of FMS technology to sheet metal fabrication. By producing over 300 parts, a true flexibility of production has been achieved. Parts produced are rather simple as compared to parts produced in the FMSs for aerospace application described previously. This indicates that cost-effective application of FMS technology does not depend on the complexity of the parts.

C. PHYSICAL CONFIGURATION

A schematic of the FMS physical configuration is given in Figure 2. The FMS can be divided into the following parts: C1 shearing center, line 1, line 2, line 3, line 4, and material handling system.

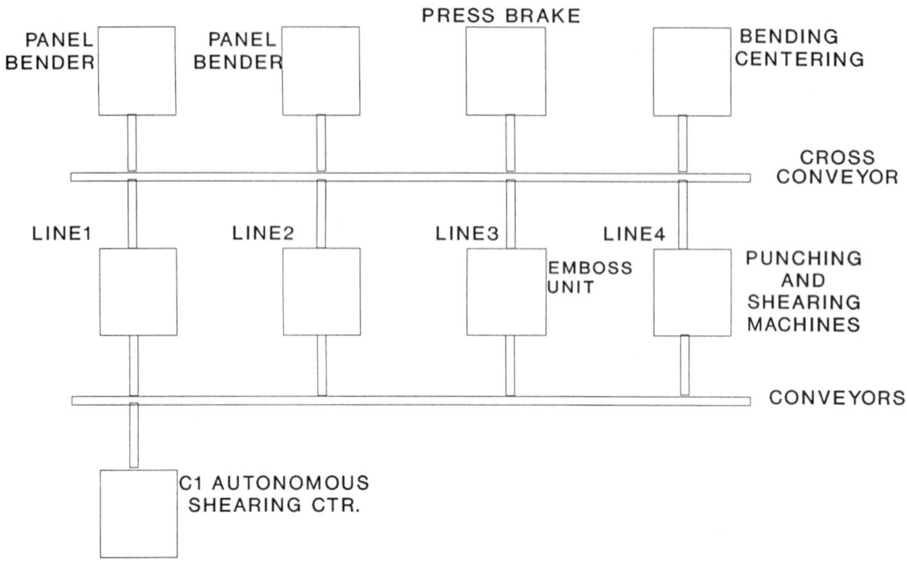

Figure 2. Flexible manufacturing system for sheet metal application—physical architecture.

1. C1 Shearing Center

This is an entry point for all the incoming sheet metal stock. The shearing center receives a daily production schedule. The sheets are loaded onto the loading tables and the large sheets are then cut into smaller sheets and transferred through a magnetic conveyor system and stacked onto a roller surface until the shear has completed packing requirements. After the stack has been completed, it is transferred to the required line. An ink jet system is used to trace the sheets. The pack distribution system consists of seven stackers or scissor tables. The stackers lift and traverse by using roller surfaces fixed to the ground. Other roller surfaces are used as buffers between the shear and individual lines.

2. Line One

As mentioned above, sheet metal stock is stacked and transferred to individual lines. A magnetic destacker is used to transfer one sheet at a time from the pack to the line. The following machines are in each line: punching and shearing machine, panel bending machine, and embossing station.

An embossing operation is required on only a few parts and hence the embossing station either performs the operation or simply transfer parts to the exit conveyor.

3. Line Two

Line two is quite similar to line one. This line has only two machines: punching and shearing, and bending. The difference between the punching operation in line one and line two is that the punching head in line two uses a special head to draw holes or both holes and knockouts. A vertical storage unit is used as a buffer between the punching and bending operations. After a final forming operation on the bending machine, sheets are transferred to the exit conveyor.

4. Line Three

Line three is used to produce many small parts with an overall volume of over 1 million per year. The shearing machine in this line is similar to the shearing machine in line two, and in addition there are five embossing units. This line has significantly more job-routing flexibility as some parts are processed through the punching machine only, while others can be processed through all three machines.

Smaller parts are transferred by an underground conveyor to a bin system that automatically sorts the parts for use in the final assembly. Some parts that require a bending operation are transferred between the lines on a cross conveyor to line two. The system also has the flexibility to transfer parts to a programmable press brake.

5. Line Four

The configuration of the punching machine is identical to that of other lines. Five double embossing units are added to the punching line. This line also includes a 100-metric-ton hydraulic press which can bend up to 32 different parts automatically.

6. Material Handling System

An automated conveyor-based elaborate system is used for material handling. These conveyors link each machine together. As described above, machines within each line are interconnected. A cross conveyor connects all four lines. The cross conveyor is used when it is necessary to transfer sheets between the lines. The cross transfer facility has increased flexibility of the system and ensured cost-effective utilization of the machinery.

Scrap removal is an integrated part of this system. Scrap removal conveyors run under all punching machines. At the shearing center there are two large scrap containers. Scrap is automatically separated into mild steel and galvanized and is stored separately into two containers.

D. CONTROL SYSTEM

As shown in Figure 3, the control system is based on a combination of a supervisory controller and a line controller. Each subsystem is described here.

1. Line Controller

Each of the four lines has a microcomputer and a machine tool controller. The machine controller directs electrical, mechanical, and hydraulic power and movements of each machine. The microcomputer acts as an interface device between the supervisory controller and the machine controller. The machine controller also monitors and reports on-line diagnostics in the form of errors and warning messages. These messages are used by equipment operators.

The microcomputer also stores master programs which control movements of each machine. This computer updates the supervisory controller on the status of each sheet metal part. This information is used to synchronize the automatic transfer of sheets between the lines. Part programs, which are essentially NC instructions for each part, are also stored at each line.

2. Supervisory Controller

This controller manages the system and communicates with the external order-entry system. Orders are transmitted from the customer order system located in Plainville, Connecticut. Orders are sorted and optimized on the same day on which they are received. Resulting production control reports and a cutting-cycle file are loaded into the shearing centerline controller the next day. The shearing center uses this cutting-cycle file to begin cutting raw stock. The raw material is stored in packs and a status message is sent to the supervisory controller by the shearing center controller. Depending on the status of a designated line, a trolley with the pack is either sent to it or removed to a waiting station.

The controller is equipped with a dynamic graphic representation system. Other special features include the ability to track scrap parts and program NC part instructions. Any scrap or rework is automatically adjusted in the next day's schedule.

E. SUMMARY

This is one of the unique applications of FMS technology. Flexibility of the system is indicated by the fact that over 300 parts are produced. The control system is adequate but less sophisticated than the control systems of many recent FMSs. The parts produced are significantly simpler as compared to parts produced in the GD FMS. But this proves that FMS implementations can be cost-effective for even simpler parts.

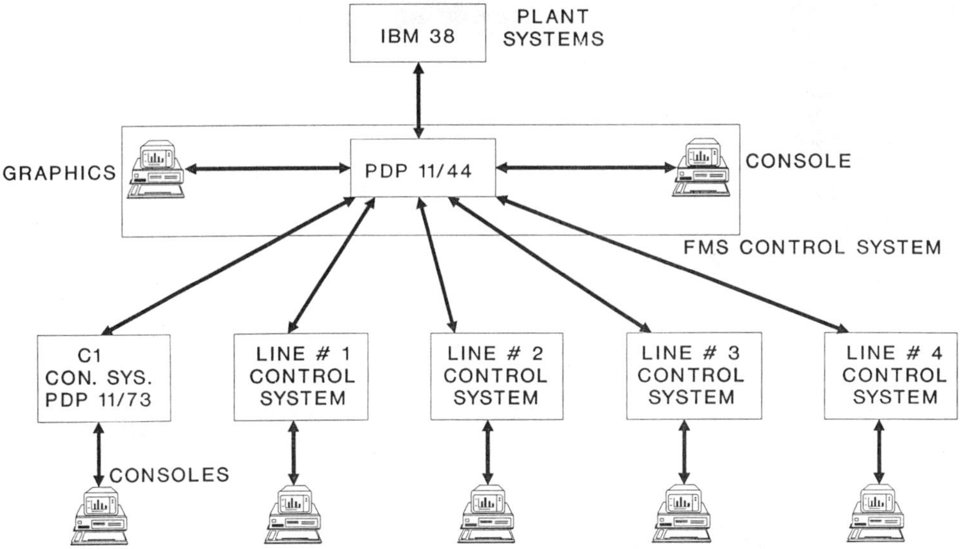

Figure 3. Flexible manufacturing system for sheet metal application—control system configuration.

V. Flexible Manufacturing Systems for Wire Harness Assembly Operation

A. INTRODUCTION

Similar to the FMS for sheet metal fabrication described above, this FMS installation shows the diversity of FMS applications in the United States. This FMS has been implemented at Westinghouse Electric Corporation facilities in Pittsburgh, Pennsylvania (Sturges *et al.*, 1984). This FMS was developed by Westinghouse along with Unimation.

It is used for the production of low-volume wire harnesses. Commercial wire stripping, termination, marking, and lacing equipment are utilized. Special robotic end effectors are used along with special wire mesh tables. A schematic of the physical configuration is shown in Figure 4.

B. PHYSICAL ARCHITECTURE

The FMS consists of five functional units; a programmable formboard, an Unimation industrial robot, end-effector tooling, commercial wire preparation and finishing equipment, and a cell control computer. This equipment is described briefly here.

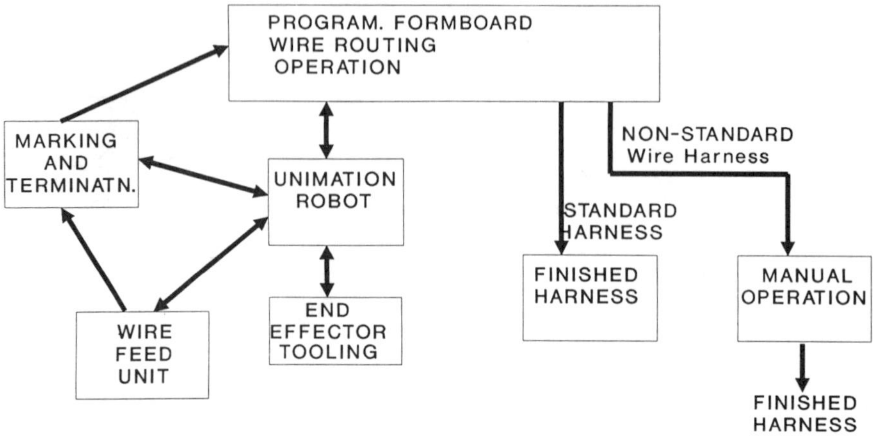

Figure 4. Flexible manufacturing system for wire harness application—physical architecture.

1. Programmable Formboard

Wire harnesses are built upon a programmable formboard. The board consists of several wire mesh layers into which reusable pins can be inserted by a robot. This formboard design has significantly improved the previous manual formboard design process. The entire table is mounted on a longitudinal trunnion which allows the table to be inverted. After a harness is completed, the formboard table is inverted. In this position, the finished harness, the pins, and other table-mounted devices can be selectively dropped off onto a conveyor underneath. To perform secondary manual tasks, the table is rotated to a vertical position. Secondary manual tasks are usually required for addition of nonstandard equipment on the harnesses.

2. Industrial Robot

A Westinghouse/Unimation 6000 robot is used to perform the manipulative tasks. The robot has five degrees of freedom, accuracy of plus or minus 0.2 mm, and repeatability of plus or minus 0.013 mm. The robot can be programmed off-line or in a teach-and-repeat mode. The robot is controlled by a Producer CNC system. The off-line programming method can utilize an APT code after postprocessing. The formboard is bolted to the base of the robot to ensure alignment between the two devices.

3. End-Effector Coupler

The diverse nature of operations requires several robot end effectors. Automated operation requires that the robot must be able to acquire these end effectors under software control. This is achieved using a coupler mechanism. The coupler is controlled by CNC external function codes. The end-effector change can be achieved in less than 10 s.

4. Wire Preparation and Finishing Equipment

This includes a wire feed unit and wire marking and termination equipment. The wire feed unit supplies wire to the robot end effectors. Five distinct wire types can be selected without a setup change. Wire-end identification is necessary in preparation of wire harnesses. Marking sleeves or tags are usually manually applied. This function is performed by the wire marking machine. Instead of sleeves or tags which require manual intervention, an ink jet marking system is used. Marking and feeding of wire occur simultaneously and hence no additional time is required. The wire termination unit acts immediately after feeding the wire. A terminal crimping machine is used. The crimper is also CNC controlled.

C. FLEXIBLE MANUFACTURING SYSTEM OPERATION

The FMS operation is based on a serial just-in-time method. This tends to reduce the handling steps even though the operation time at each step is increased. The wire feed unit CNC controls the wire to feed. The feed unit then selects the next wire while the robot moves a measured length. This is called a feed-and-selection cycle. In this cycle all wires are positioned at the same location in the feed cycle for dispensing, marking, and cutoff. Prior to cutoff, the marking unit applies identification data on the wires. At this time, the FMS terminates one end of the wire before routing. A wire-routing end effector is coupled to the robot for the wire acquisition, terminating, and routing tasks. The routing is done on the wire formboard described above. After routing, any nonstandard wires or terminals are applied manually by the operator. Lacing is performed by FMS at this stage. The wire harness is complete and ready for removal after the lacing operation. The harness removal can be manual or by using a harness-remover end-effector on the robot.

D. SUMMARY

There are very few industrial products which can be manufactured using purely machining processes. A large portion of the manufacturing cost is associated with processes like welding, casting, forging, assembly, and heat treatment. This FMS shows that the principle of flexible automation applies equally to machining and nonmachining processes.

VI. Conclusion

The above survey clearly indicates the variety and complexity of FMS usage in the United States. In earlier years, the developments in FMS area were focused only toward manufacturing improvements.

The U.S. FMS technology has always been considered world class, but U.S. management has been criticized in the past as unwilling to use the full potential of FMS technology (Jaikumar, 1986).

Recently, there has been a change in the perspective and manufacturing is considered not as a liability but as an asset of strategic importance. As manufacturing is regaining its rightful place in the U.S. business environment, the FMS technology is also being considered as strategically important to business success. Specifically, FMS technology is considered as essential for achieving product differentiation, product and process innovation, and customer satisfaction. The FMSs are also considered valuable in a high-volume environment, especially to manage rapid product enhancements (Venkatesh, 1990). More attention is being devoted to human resource issues in this field (Nemetz, 1990).

The developments in the next 5 to 10 years will determine if the United States continues to be a leader in FMS technology, which is vital to any industrialized society.

References

Dutton, B. (1989). Cat climbs high with FMS. *Manuf. Sys.* November, pp. 16–21.

Jaikumar, R. (1986). Postindustrial manufacturing. *Harv. Bus. Rev.* November–December, pp. 69–76.

Klahorst, T. (1984). FMS. *In* "Flexible Manufacturing Systems" (J. R. Holland, ed.), pp. 3–6. Society of Manufacturing Engineers, Dearborn, Michigan.

Nemetz, P. L. (1990). Human resource planning in flexible manufacturing organizations: Possibilities and concerns. *Manuf. Rev.* **3,** 16–22.

Palframan, D. (1987). FMS: Too much, too soon. *Manuf. Eng.* March, pp. 34–38.

Parker, G. M. (1987). Implementation of an advanced FMS at General Dynamics. *CASA/SME Flexible Manuf. Sys. Conf., 1987.*

Runewicz, D. (1987). An FMS for the production of sheetmetal panelboards. *Proc. Int. Conf. Flexible Manuf. Sys. 6th, 1987,* pp. 279–288.

Sturges, R. H., Jr., and Wingert, C. D. (1984). *Wire Harness Flexible Manuf. Sys., Autofact 6 Conf.*, pp. 16-1–16-20, Anaheim, California, October 1984.

Venkatesh, R. (1990). Cummins engine flexes its factory. *Harv. Bus. Rev.* May–June, pp. 120–127.

CHAPTER 10

Factory of the Future

H. J. Warnecke
University of Stuttgart and
Fraunhofer Institute for Manufacturing
Engineering and Automation (IPA)

I. Introduction

The development and application of flexibly automated manufacturing equipment, allowing for the complete automation of almost all main and secondary functions even for widespread production tasks, has shown substantial progress in recent years.

The first flexible manufacturing systems, established in the 1970s, were mainly designed for machining prismatic workpieces. Development concentrated on the machining and transport equipment. Flexibility, however, could not be achieved solely by the optimization of single elements, but required the simultaneous consideration of all components. This insight led to a huge demand in planning facilities for FMS. The complex task required planning techniques that considered the whole system. A new research field arose, devoted to the development of methodical FMS planning aids such as data bank-supported tools for machining and material flow equipment, as well as simulation techniques. Manufacturers and research institutions such as the Fraunhofer Institute for Manufacturing Engineering and Automation (IPA) established working groups whose experiences reflect the development of the state of the art in flexible automation (Steinhilper *et al.*, 1988).

II. Main Developments in Flexible Manufacturing Systems toward the Factory of the Future

The increasing importance of information flow planning and development is shown in Figure 1. It is likely that in a few years half of the work involved with FMS projects as a whole will be taken up with this aspect. This is in line with findings by the consultancy company, McKinsey, which forecasts a 50% software component in FMS, making this sector the decisive market segment in factory automation of the 1990s.

The demand for development resources aimed at FMS elements will not decrease in the future. Many FMS components are commercially available now,

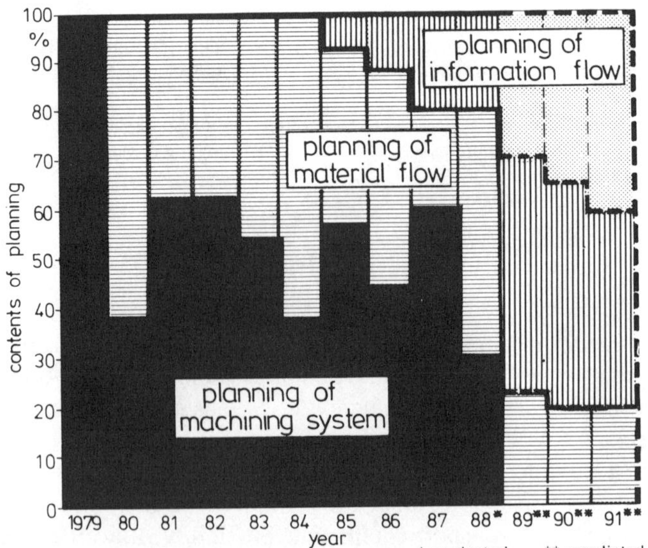

Figure 1. Development of contents in FMS projects at IPA.

so that one might expect the main task of an FMS project to consist of the accurate planning and arrangement of known components. More and more development tasks for FMS are still arising, however. The implementation of software in particular remains a major problem. Standardization of interfaces is one way of reducing the expense for adaptation. The main task, however, is the reduction of the number of interfaces.

It is not only parts, and possibly tools, which need to be transported in flexible manufacturing systems. The automated clearance and disposal of chips from machining is a task which did not occur on working schedules previously and which was often left out of consideration until the automation stage. Here many detail problems have to be solved. A single chip at the wrong spot can disrupt automated machining and material flow and can bring the whole system to a standstill. Attention to supposedly unimportant things such as chip flow and the provision of suitable technical and organizational measures to deal with them may be key factors for the failure-free operation and running of the material flow in future FMS.

Not only the project contents but also the products to be manufactured in FMS have changed. Since 1980, development efforts have been aimed at the design of FMS for turning, welding, and sheet metal working (Figure 2). The special problems of sheet metal working with regard to flexible manufacturing have resulted in an average difference of one generation between part manufacturing and sheet metal working. There is thus a great potential for improvement.

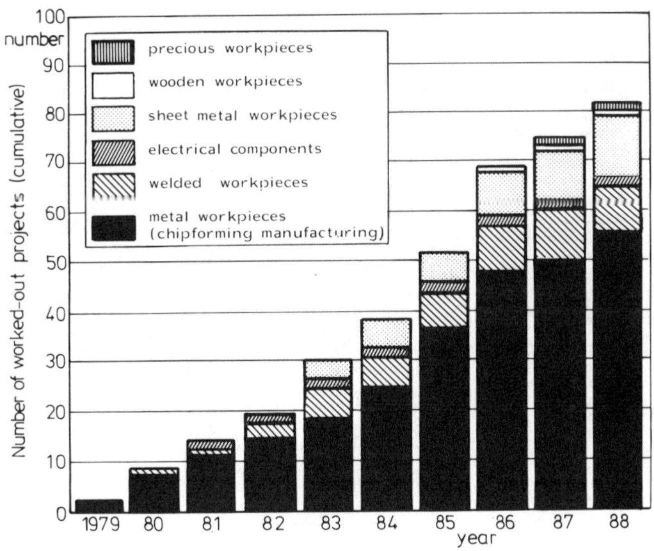

Figure 2. Types of workpieces in FMS projects (based on IPA studies).

A. FLEXIBLE SHEET METAL FORMING

The restructuring of sheet metal part production requires a methodical approach. Figure 3 shows the flowchart for computer-aided planning of a sheet metal part production system. Using the overall workpiece spectrum in current production, together with the suitable workpiece data, a representative workpiece spectrum is established for the planning period in question. Here future market developments may have some influence. From this representative workpiece spectrum individual samples are selected, with the aid of which the parameters for the design of the new production structures can be calculated.

Only with this computer-aided planning will it be feasible to detail the product requirement specifications to the required accuracy.

The procedure for planning a new production structure and the results achieved can be illustrated by an example for the cutting of sheet metal parts (Warnecke and Claussen, 1988). The objectives were

- Restructuring of technologically high-grade sheet metal parts production
- Upgrading of automation
- Tightening of the material flow
- Increase in productivity and economy in sheet metal parts production

Figure 4 shows different stages of automation for the production of sheet metal cutting. On the left, manual loading and unloading can be seen, whereas on the right of the picture the level of automotion is up to fully automatic cutting centers with unstacking and sorting facilities. The production concept selected in this

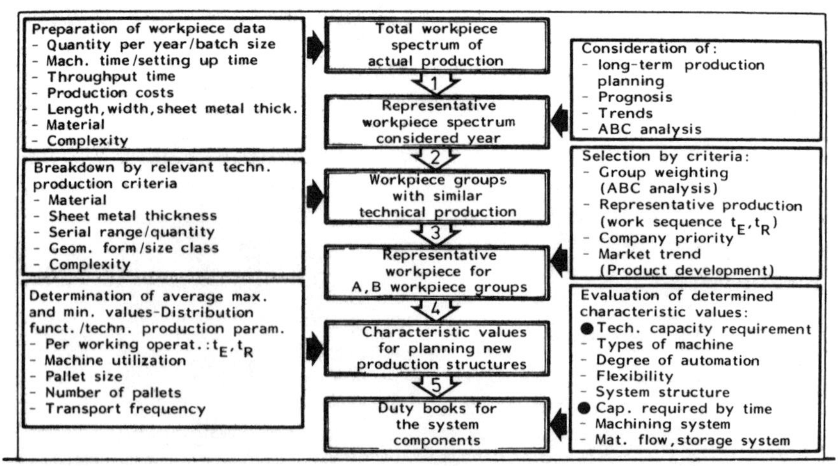

Figure 3. Computer-assisted planning of a sheet metal part production system.

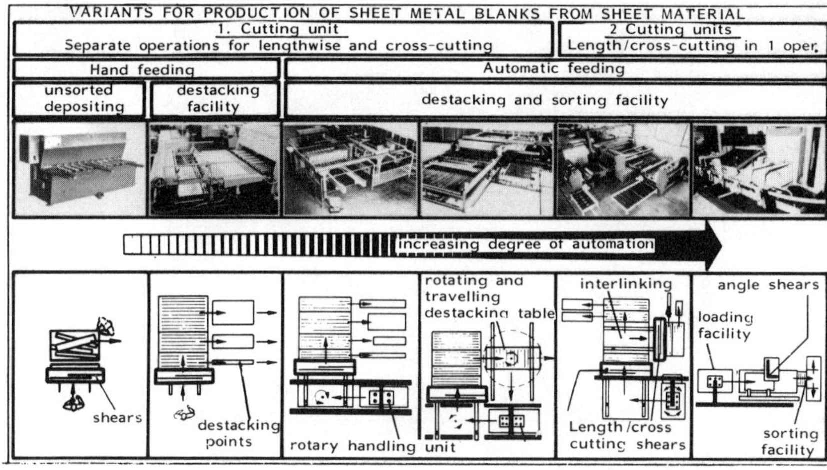

Figure 4. Stages in automation for production of sheet metal blanks.

project consists of a CNC cutting center with an automatic loading and unloading device, which provides the most effective rationalization. The selection of the new concept is based on economic feasibility. This factor can be determined with sufficient accuracy by the planning method described here. The investment totals just over DM 0.5 million, resulting in an amortization period of under three years. In this calculation, only directly ascertainable factors were taken into account. The indirect benefits such as improved quality, increased flexibility, stock reduction, and so on must not be overlooked in the evaluation.

This automated cutting section is, as described, an automation island. To arrive at really new production structures it must be integrated into a flexible automated overall concept. The result of this planning measure can be seen in the layout (Figure 5). The individual areas are interlinked via an automatic sheet metal rack storage system. This direct linking permits a very flexible automatic production sequence, even with the extremely heterogeneous workpiece spectrum. Compared with the initial condition, the production structure is much clearer.

The machining processes in conventional sheet metal production structures are highly developed. The setup and downtimes, however, take up a much larger portion of the time required for job processing. Multiphase production sequences with high setup time quotas and a low level of automation result in long throughput times and high intermediate stock levels. To avoid the disadvantages of conventional production structures and to achieve increased productivity, economy, and flexibility in the sheet metal parts production overall, a number of objectives must be pursued in the future:

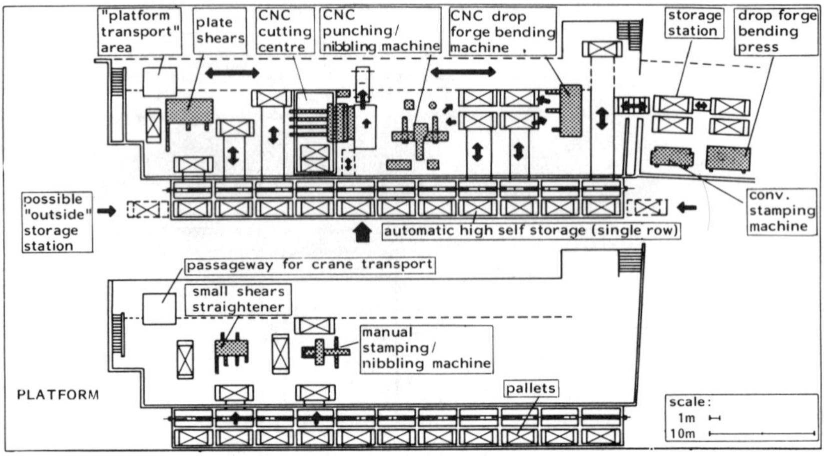

Figure 5. Incorporating new machine concepts.

• New or alternative processes or process combinations, for example, combined punching and laser cutting instead of complete cutting

• Creation of flexible production cells with automatic loading, machining, unloading, and monitoring

• Suitable organization and production control structures

One possible solution, the result of another project, provides for two flexible CNC-controlled nibbling machines directly interlinked with an automatic high-bay storage system. The unloading of finished parts and scrap, and the tool change, are automatic. In considering the investment cost of such a production system, it is remarkable that the basic machine costs represent only 40% of the total investment. Automation components, including storage, account for nearly half the total cost (Figure 6). The results achieved with this system are encouraging. With an investment total of DM 3 million, an amortization period of 2.7 yr is achieved; the throughput time can be reduced by 20% and the unit price by 22%. Apart from economic feasibility studies, other advantages need to be taken into account in the overall assessment (Figure 7).

The planning of an information flow system must be based on the analysis and structuring of the desired production. Only by this process is a realistic evaluation and subsequent hardware and software implementation for computer-controlled production possible.

The basis for the compilation of an information flow system is the establishment of the future production structure. The performance potential of such production systems can be fully explored by the user only if the production

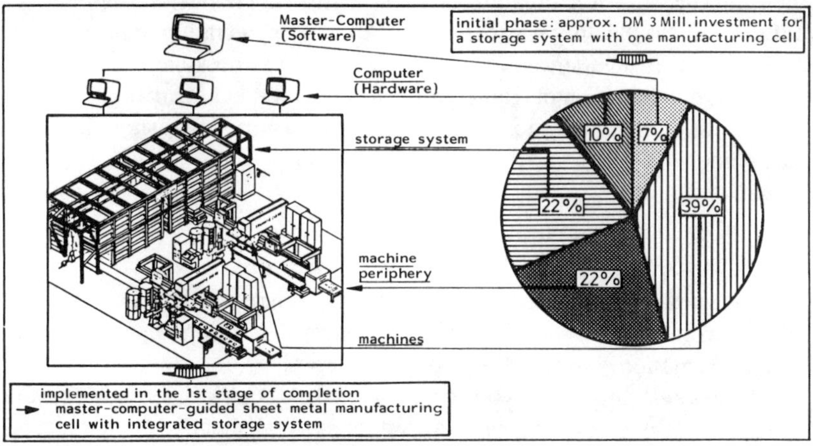

Figure 6. Quotas at the investment volume of an FMS.

sequence is planned, arranged, initiated, controlled, and fed back by a computer-controlled production control system.

Due to the different sequences within the system, it appears expedient to create organizational subsystems with specific limits that can be integrated as modules into the overall system. The subsystems are defined in accordance with the functions performed within their system limits and the objects interchanged

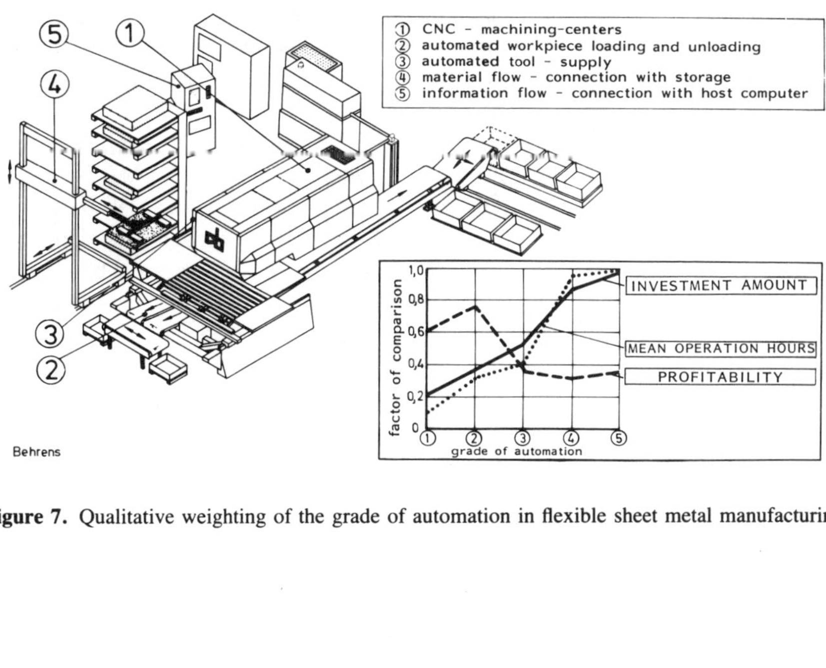

Figure 7. Qualitative weighting of the grade of automation in flexible sheet metal manufacturing.

at their interfaces. In the creation of subsystems, functional overlapping is to be avoided, as this is the only way that a "blackbox" consideration of the subsystems, irrespective of the sequences in other subsystems, is possible.

In the development of framework concepts for the information flow system, the possibilities of integration of further computer modules should be investigated, namely

- Factory data acquisition systems
- Production planning systems
- Quality assurance systems

The variants developed are evaluated by criteria established in close cooperation with all partners involved. The selected solution is specified and the system introduction described in a step-by-step plan.

Only when this entire volume of work has been carried out can the appropriate implementation of such a system be realistically evaluated and the implementation planning initiated.

B. SIDE EFFECTS OF FLEXIBLE MANUFACTURING SYSTEM PLANNING PROJECTS

Important side effects of the planning and installation of flexible manufacturing systems are not only the big increase in the company's know-how in manufacturing technology and all connected tasks, but also quite a few previously unexpected benefits and surprises, which can appear together with an FMS installation or as an alternative to its implementation. During the planning and preparation of its application, several operation areas require thorough scrutinizing and alterations. During this process further weaknesses are identified, which can result in the reexamination of other principles hitherto regarded as incontestable.

For example, a thorough check of the product line will certainly be made and machining processes will be altered, whether or not the FMS is installed later. If an installation follows a planning procedure carried out carefully enough, the biggest "surprise" is often the unexpected rise in capacity, which occurs especially in order-oriented small- and medium-batch production. The operation of an FMS, or in some cases even the planning alone, also successfully helps to reduce many presently "hidden" costs of manufacturing.

It is an almost impossible task, however, to accurately quantify all the additional expenses and problems occurring, especially in conventional medium-batch production, in terms of pure planning and controlling costs of the production, tooling, jig, setting up, maintenance, scrap, reworking, and all the other possible costs and to apportion them to the correct cost centers.

Here lies the secret of the success of flexible manufacturing systems. The reduction in the hidden costs gives much higher potential for rationalization and saving than the reduction in direct labor costs, which is usually the aim of automation.

C. SUCCESS-DETERMINING FACTORS DURING FLEXIBLE MANUFACTURING SYSTEM PLANNING AND DEVELOPMENT

Flexible manufacturing systems are not just a new technology—their planning and introduction onto the shop floor also signifies a new way of thinking in many areas throughout the company. Therefore, the most important success-determining factors during FMS planning and development cannot be described by technological data or performance. Experience with many planning procedures and successfully operating FMSs shows that the key to success lies mainly in the attitude toward FMSs of the people involved in planning, design, installation, and operation. Only where the project team's members completely identify themselves with the task, where there is teamwork and the ability for cooperation and coordination within all phases of the project and between all participants, can successfully operating equipment and satisfactory figures result.

The FMS can still not be regarded as a turnkey solution, and will still call for a pioneering spirit and the willingness to face changes that occur not only in the manufacturing shop but in many departments and at all stages of order handling. Especially on the information flow side, project teams who look intensively for an adequate and accurate supplier of the controls and software will have much less success than those who agree to contribute themselves to the new developments and modules necessary to run a complex system carefully tailored to the particular needs of their company. They have to be prepared to acquire a large amount of new knowledge and know-how which cannot just be bought from a supplier, but will emerge during planning, installation, and operation of the system (Figure 8).

Of course, carefulness in selection of the right suppliers of machining, material flow, and information flow equipment as well as security and accuracy at all stages of the planning procedure are just as important as teamwork and joint responsibility. The latter especially decisively affects the correctness of the data input, such as production batch sizes, machining and quality requirements, and working sequences of the workpieces into the design and dimensioning of the system. Whereas conceptional planning and detailed planning can be done mainly by a consultant, this data input and its accuracy will always be the task and remain the responsibility of the customer.

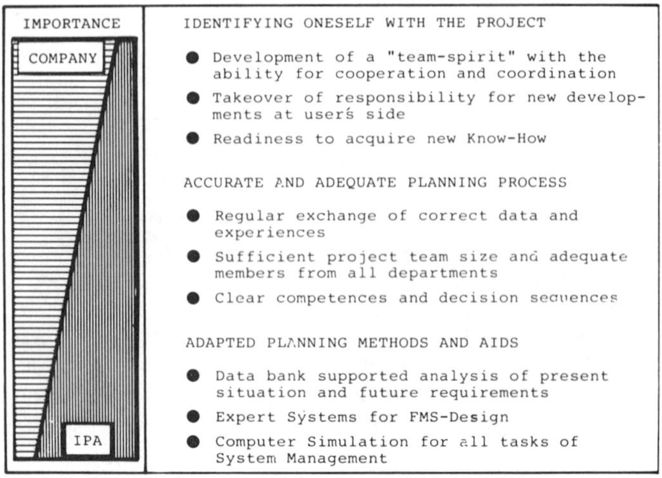

Figure 8. Success-determining factors during FMS planning.

III. Artificial Intelligence and Expert Systems in Flexible Manufacturing Systems

A. COMPUTER-AIDED PLANNING AS A CENTRAL PLANNING AREA

Since essential tasks in the conceptual and operative phases of the manufacturing system are the responsibility of operations planning, considerable interaction can be noted, especially between process planning and the economical introduction of flexible manufacturing systems. Thus operations planning determines the investment costs for processing and material flow facilities, as well as manufacturing system operating costs (Roth *et al.*, 1988).

During the planning and configuring of flexible manufacturing systems, work plans are important for analysis of the current status and the future workpiece spectrum. As such they form the basis for new manufacturing concepts (Figure 9).

In addition to the representative workpiece spectrum, the process plans must be prepared in the initiation phase for all workpieces to be processed in the flexible manufacturing system. Also, a multitude of special requirements must be considered in respect of the manageable complexity of workpieces or machine tool process reliability, with low staffing levels.

A process plan update is often carried out during adaptation of construction or process times on new manufacturing concepts with higher productivity. A

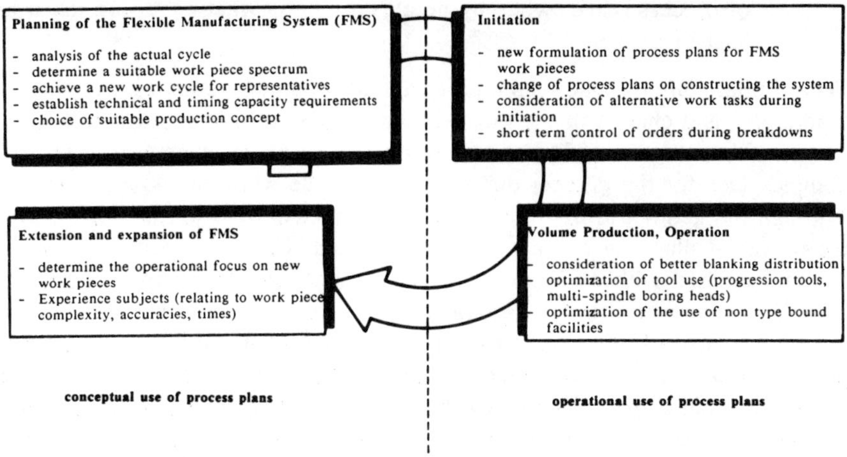

Figure 9. Use of process plans in the different phases of an FMS.

fundamental consideration of changed chucking situations or of a changed tool requirement occurs in very few cases.

In the course of planning activities for a flexible manufacturing system, as the long-term task of operations planning, it is necessary to retain detailed knowledge obtained about manufacturing possibilities.

B. BASIC TASKS AND METHODS OF PROCESS PLANNING

Within the production development process, operations planning represents the link between product definition in design and product realization in manufacture.

Computer aids can be classified into the following four basic types:

1. *Repeat planning*: The simplest type of process planning, slight or form changes on plans already existing. However, no new planning data prepared.
2. *Variation planning*: Already existing standard process plans are used as a basis for planning workpieces within corresponding part families and planned from the resulting geometry or quantity variations.
3. *Similar planning*: By means of the plan of a workpiece with similar geometry and manufacturing technology, planning takes place by change and adaptation of individual process operations. Finding a similar part presents an essential problem.
4. *New planning*: This method is applied to a basically new preparation

of process plans, whereby no existing parts whatsoever are taken as a basis.

With similar planning, the search for an already planned, geometrically classified workpiece, and change of the most similar process plan is possible to a limited extent with computer aid. New planning not based on a pattern is the most complex task for the planner due to the use of the whole workpiece spectrum. Through the geometric multiplicity generally present in a production program, the universal planning process is not automated. However, part functions can be implemented with computer aid. The organizational principle of operation planning represents a further significant parameter in a computer-aided system.

Here two different principles have established themselves in operational practice. In object-oriented process planning, the planner prepares the complete process plan of a workpiece consistently through all planning functions. He or she must possess a comprehensive and detailed knowledge of procedures and machines. Function-orientated planning requires detailing of the individual process operations by specialists, after rough planning of the workpiece has been carried out in advance in a central location.

This organization principle can also be broken down into subgroups—procedure-specified (turning planner, milling planner) or technology-specific (NC planner, planner for flexible manufacturing system).

C. INTEGRATION INTO THE INTERNAL INFORMATION FLOW CONCEPT

The efficiency of a manufacturing area, and particularly the means of manufacture, has increased considerably in recent years. By contrast, the immediately adjacent areas have been largely excluded from this development despite the considerable contribution they make to the overall costs. In this respect, it would appear that too much emphasis has been placed on rationalization and automation of production as a way of optimizing the manufacturing process as a whole.

While CAD systems and databanks with subject characteristics are well established, computer-assisted process planning is still in its infancy.

It would be expedient in design, for example, to be able to fall back on a module of a CAP system, and thereby to make possible a cost estimate of alternative designs. This problem arises from the fact that considerable changes in the work cycle result from slight geometric differences, such as a reduction in wall thickness between two pockets, difficult to achieve without explicitly preparing a new process plan. By contrast, a basic feasibility check of the design drawings represents a soluble problem for a software system. The production planning and control system, in which the order data are managed after transferring the part master files to the planning system and the resulting prepared process plan, must be automatically available for timing and capacity planning.

In this context, especially with reference to short-term preparation of process plans, it is desirable to be able to react to breakdowns in the workshop, and also in exceptional situations to adopt the most economical solution—hence the requirement to be able to introduce components from CAP systems singly. In practice, when a processing center fails and is subject to timing deadline pressure, a change tends to be made to universal boring and milling machines without considering available facilities and tools for reducing chucking.

The sought-after goal, however unachievable at this point, is to couple the planning system to the industrial management and technical areas of a company, for which the process plan or the information contained therein must be available.

D. KNOWLEDGE-BASED PLANNING AIDS

Different degrees of automation support are possible, depending on company-specific conditions, which can be implemented with various software systems (Eversheim and Schulz, 1985).

1. Decision Table Systems

Decision table systems represent a preliminary stage in the direction of automatic systems for process plan preparation. As an aid, they offer procedures from manual dialog-oriented plan preparation, through semiautomatic functions, to fully automatic process plan generation. By means of decision table technology, intelligent dialog can be achieved with reduced input expense. Above all, however, projection of the process plan logic contributes to raw material operations, machine and processing stock determination, and time study. Furthermore, the required hierarchical construction of the system contributes considerably to the systemization of planning activities. With the aid of decision table generating and interpreting programs, the tables can be generated and modified relatively easily.

Decision tables are to be considered as a form of knowledge-based system, in which, compared with conventional programs, company-specific procedures can be represented relatively simply for problem solving in a rule-oriented manner. In the ideal case, the construction of these systems should be carried out by the users themselves, using the available internal system aids provided. In practice, however, the manufacturer's planning experience is required, above all in difficult knowledge structuring.

2. Expert Systems

In comparison to conventional systems and as an extension of decision table systems, expert systems make it possible to process planning knowledge in the form of rules and facts. Knowledge is no longer in firmly programmed or tabular form, but given to the knowledge base without a rigid structure. The finishing of this planning knowledge takes place through a so-called inference component, which deduces results for a specific application from the user's input.

In recent years, process plan preparation has been tackled intensively with the help of expert systems, especially by universities. However, at this point only a few implementable formulations are known. They are nevertheless promising, particularly in the area of rotational workpieces.

Expert systems are ideal for planning tasks in particular, not least as a result of the planner's method of working, which involves breaking down a problem initially into a number of subtasks and devising partial solutions for each of these tasks.

When all the alternatives are considered together, a large number of combinations present themselves. All-purpose software aids are incapable of handling this profusion of concepts, and planners are obliged to resort to knowledge-based problem-solving techniques.

From the planner's procedures in new planning, a model can be defined consisting of five planning steps (Figure 10). This is valid for planning the total part spectrum, with individual workpieces run through repeatedly.

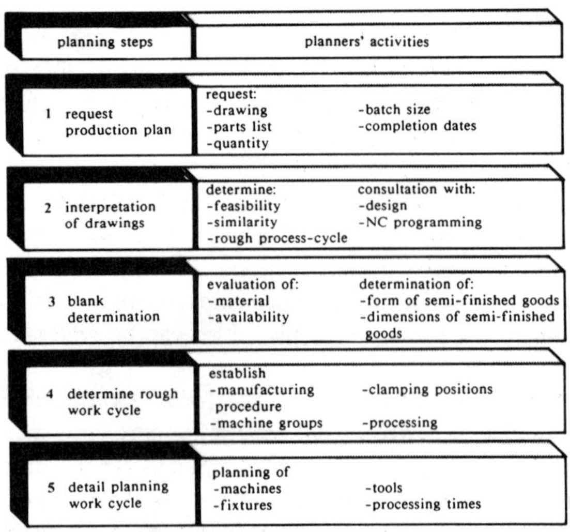

Figure 10. Planning procedure phases in the preparation of the process plan.

Construction of the expert system, itself composed of dialog and inference components and a knowledge base, is identified through the tie-up to the CAD system and the knowledge base divided up in different modules (Figure 11). The dialog component represents the interface to the user not only in the actual course of planning, but supports the user as an aid in form recognition. This calls for an intrinsic and discriminating inference process in the planning segment.

The usefulness of expert systems in practice depends essentially on the construction and content of the knowledge base. One of the main problems in setting up a system is therefore to ensure transfer of knowledge from experts to the users. Since a system for process plan preparation must be constructed over several years, and availability of experts is as a rule limited, the development of a suitable procedure for knowledge acquisition should not be underestimated.

At the start of the project a current analysis of organization, methods, and procedures in the process planning is indispensable, since the introduction of an expert system for process plan preparation is not expedient in every case. For planning methods based on a rigid sequence or where computer-assisted planning is available only for certain subtasks, the use of conventional systems is more suitable (Figure 12).

As an extension to the spectrum of systems for computer-aided process plan preparation, expert systems offer a very good specification for future implementation due to the application potential of this new technology.

Computer-assisted planning is only one potential application for expert systems. Intensive research aims at the development of software for FMS design

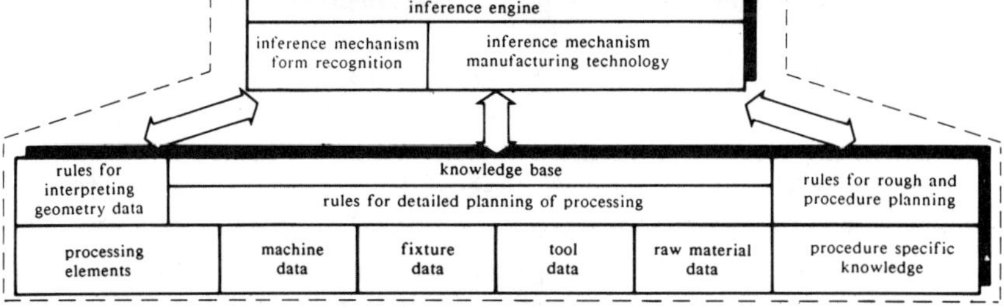

Figure 11. Structure of an expert system for process planning.

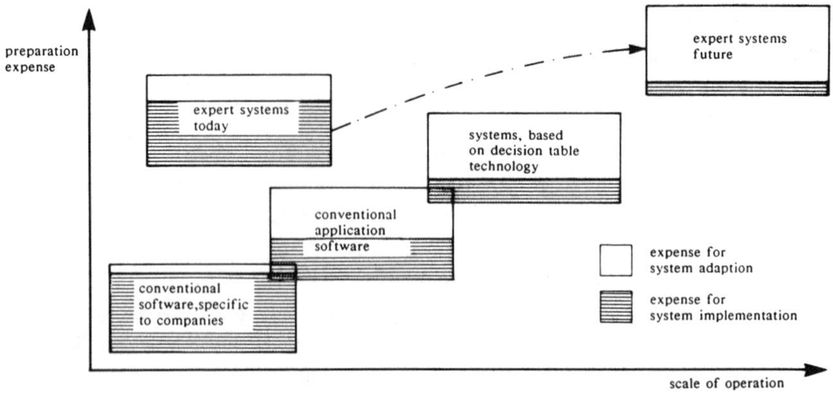

Figure 12. Different CAP systems.

and connected problems. Currently, however, only prototypes with restricted capabilities are known.

IV. Design Philosophy and Characteristics of Future Machine Tools in Flexible Manufacturing Systems

A. MACHINE TOOLS IN FLEXIBLE MANUFACTURING SYSTEMS

What will be the main characteristics of flexible manufacturing systems in the 1990s? A survey recently carried out by the Fraunhofer Institute for Manufacturing Engineering and Automation on FMSs already partially installed, planned in detail, or designed conceptually for the 1990s showed that an impressive range of all kinds and sizes of FMS is planned in Germany in the coming years (Figure 13).

Although lathes and machine centers will still remain the kind of machine tools used most frequently in future FMSs, one-third also involve grinding machines and nearly one-half contain other machine tools (Figure 14).

Looking at additional operations to be included into FMS, Figure 15 shows that there is a great need for "system machines of the next generation" which should also be available for non-chip-forming operations and adaptable to the material flow and information flow interfaces within FMS.

Efforts are being made in the development of system to further automate tool handling in FMS (Figure 16). Even now the technology exists to deliver

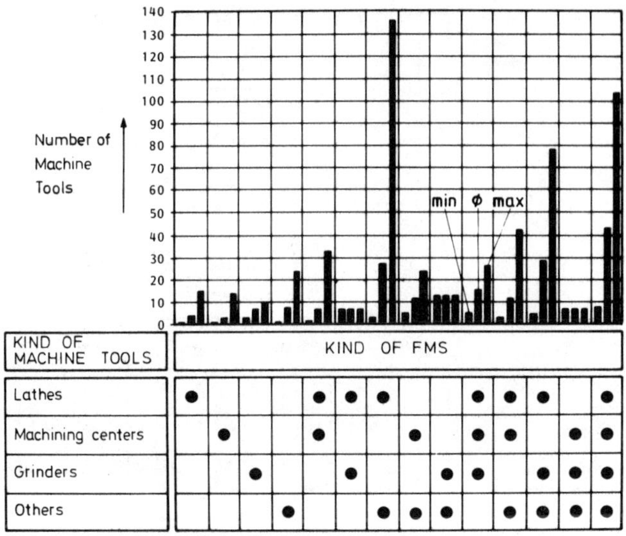

Figure 13. FMS of the 1990s in Germany: number of machine tools.

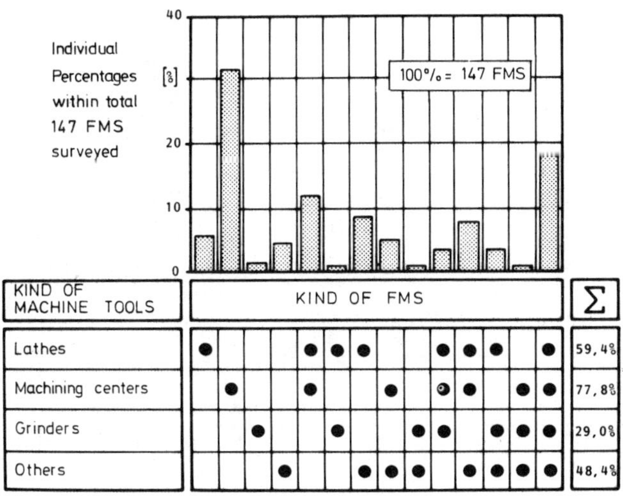

Figure 14. FMS of the 1990s in Germany: variety of machine tools.

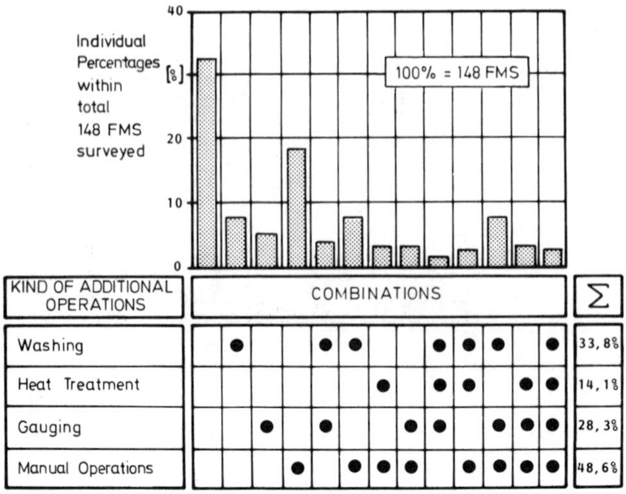

Figure 15. FMS of the 1990s in Germany: integration of non-chip-forming operations.

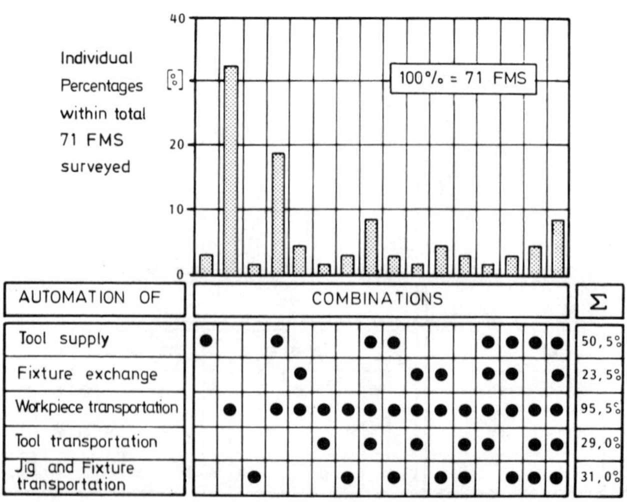

Figure 16. FMS of the 1990s in Germany: material flow.

tools within the FMS using mobile robots and AGVs. Sensor technology and related algorithms will progress. Thus it will be possible to predict tool wear and tool breakage more accurately. Inspection of workpieces will also be a target for FMS suppliers as links between coordinate measuring machines and FMS control are refined and fully exploited.

Some experts predict breakthroughs allowing more inspection to be performed on the machine tools themselves. Advances in adaptive control will continue. Automated compensation at the machine tool for drive axis error and the effect of misalignment forces, workpiece weight, temperature, and machine component wear will also improve.

The link between CAD and CAM will be increasingly exploited. Designers will create families of parts designed specifically for efficient machining using FMS technology. The future will also bring new sophistication in FMS operating software. Start-up procedures will be easier and less time-consuming. Much work will be done to automate fixture storage and retrieval systems to lessen the work of setup and tear-down.

Further advances will emerge in serialization. Every tool, fixture, pallet, cart, or other transportable element of an FMS can be identified and tracked for improved overall system control (Figure 17). Application of bar-code scanners and other identification technologies will increase.

To date, most machine tools and their peripherals have not been prepared for these purposes. For this reason, presently FMSs contain a multitude of

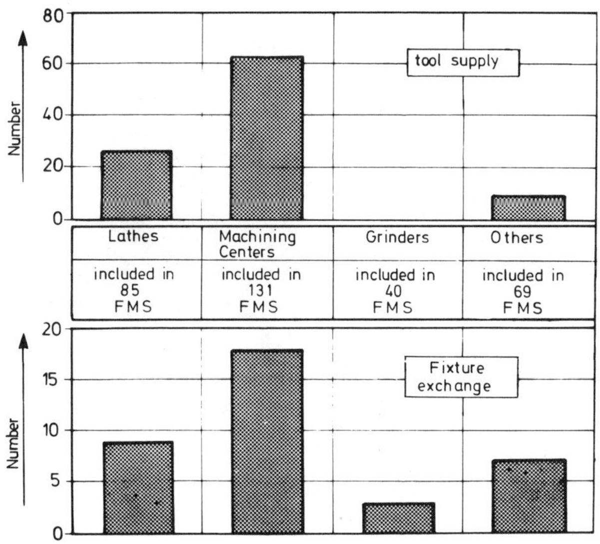

Figure 17. FMS of the 1990s in Germany: tool supply and fixture exchange.

different workpiece and tool carriers, pallets, and other handling devices which considerably complicate investment to a hardly affordable amount. "System machines of the next generation" should therefore use common standards and interfaces not only for material flow, but also, and not less important, for information flow.

B. MODULAR DESIGN OF FLEXIBLE MANUFACTURING SYSTEM CONTROL SOFTWARE

The key to flexibility is software. Controls and software can represent as much as one-third to one-half of the total investment of a complex system. Because of this, and to provide a "learning-by-doing" possibility in a step-by-step approach from automation islands to integrated systems, modular design of software packages will be absolutely indispensable. These packages will be assembled in different combinations, according to different individual applications and needs. Figure 18 shows the main goals stated in the survey mentioned above and the resulting requirements for FMS control.

Furthermore, FMS controls and software have to be compatible with business information systems, so that they can successfully interact with the overall system using a common database for design, manufacturing, and all other tasks within the entire operation, thus becoming a step toward and a part of computer-integrated manufacturing (CIM) in the automated factory of the future.

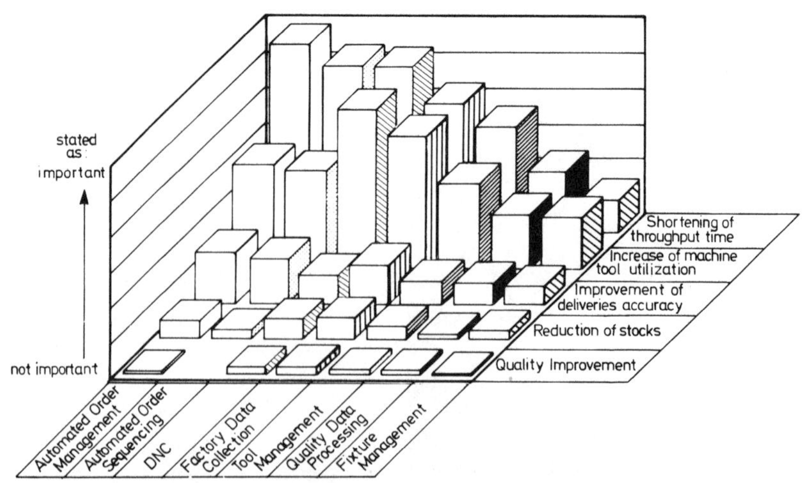

Figure 18. FMS of the 1990s in Germany: main goals and requirements for FMS control.

Since there is no standard factory, effective integration concepts can be created only on the basis of exact knowledge of the production sequences, the operational sequences, and the necessary information architecture of the participating departments.

To facilitate this overall concept, some planning work will initially be necessary, since CIM is primarily concerned with planning and organization.

The technical obstacles lie in the lack of compatibility between computers and software. Interface standards are, in this case, the key elements for the future spread of CIM applications. It will certainly be some time, however, before all those involved in the decision-making process have overcome their reservations with regard to CIM. In this respect it should be noted that time-critical processes all pose greater problems than mere MAP. The absence of coordination between existing system software, bought-in application software from different sources, and in-house software obstructs integration and productivity of the EDP service. The personnel shortage will further increase the demands on suppliers of system software with respect to training and support. Also lacking is modular software which can be distributed with little adaptation for various applications—and thus more cheaply for the individual. Fifty percent "standard," the "remainder" adaptation—these are rather optimistic hopes and estimates. Practice generally shows that a genuinely tailor-made solution must be developed from scratch and also that it will never be really complete. Computer-integrated manufacturing involves enough changes within a company so that, as a rule, the software should be adapted to the organization in the company and not vice versa. This means, however, that even the mere specification of a complex solution by a potential supplier will have to be paid for.

C. FUTURE COOPERATION IN THE FACTORY AUTOMATION MARKET

The most frequently asked question is, "What kind of general supplier should one turn to for an overall factory automation concept?" The manufacturer of existing commercial data processing systems? The CNC and DNC machine tool supplier? Or an independent systems and software house? The choice is not easy—the leap from hitherto isolated computer applications to CIM is as great as that from a company car to a company plane, something which would certainly not be ordered from the former car supplier (although there are companies who make both cars and planes!).

The question will probably stay unanswered in the next decade. It might be important, however, to remember that there are two essential prerequisites for successful extending right down to the factory floor: a sophisticated software concept and the skilled knowledge of production procedures within the company and on the shop floor itself. The latter at least should be provided by the user company at the planning stage, no matter whether a general supplier for the integrated overall solution or various component manufacturers are involved in the project. This entails familiarity with all specific components, especially with "system machines of the next generation," which in spite of all CIM discussions will continue to play an important role in the success of manufacturing companies.

References

Eversheim, W. and Schulz, J. (1985). Survey of computer-aided process planning systems. *CIRP Work. Seminar Comput.-Aided Process Plann. Syst., 1st, 1985,* pp. 1–8.

Roth, H. P., Zeh, K. P., and Muthsam, H. (1988). Artificial intelligence in future manufacturing. *Int. Conf. Flexible Manuf. Syst. 7th,* Stuttgart, *1988,* pp. 249–263.

Steinhilper, R., Storn, H., and Reinhard, T. (1988). A message from FMS work during 10 years. *Int. Conf. Flexible Manuf. Syst. 7th,* Stuttgart, *1988,* pp. 3–12.

Warnecke, H. J., and Claussen, C. (1988). Planning of flexible sheet metal manufacturing systems. *Int. Conf. Flexible Manuf. Syst. 7th,* Stuttgart, *1988,* pp. 301–315.

Index

B

C

D

G

H

L

M

N

O

T